Narrative Podcasting in an Age of Obsession

Narrative Podcasting in an Age of Obsession

NEIL VERMA

UNIVERSITY OF MICHIGAN PRESS
ANN ARBOR

Published in the United States of America by the
University of Michigan Press
Manufactured in the United States of America
Printed on acid-free paper
First published June 2024

A CIP catalog record for this book is available from the British Library.

Library of Congress Cataloging-in-Publication Data

Names: Verma, Neil, author.
Title: Narrative podcasting in an age of obsession / Neil Verma.
Description: Ann Arbor : University of Michigan Press, 2024. | Includes bibliographical references (pages 223–243) and index.
Identifiers: LCCN 2024000499 (print) | LCCN 2024000500 (ebook) | ISBN 9780472075218 (hardcover) | ISBN 9780472055210 (paperback) | ISBN 9780472129881 (ebook)
Subjects: LCSH: Podcasts—History and criticism. | Podcasts—Social aspects. | Podcasting—History. | Podcasting—Social aspects. | Social media addiction.
Classification: LCC PN4567.7 .V47 2024 (print) | LCC PN4567.7 (ebook) | DDC 791.4609—dc23/eng/20240214
LC record available at https://lccn.loc.gov/2024000499
LC ebook record available at https://lccn.loc.gov/2024000500

Contents

Digital materials related to this title can be found on the Fulcrum platform via the following citable URL: https://doi.org/10.3998/mpub.11751593

Figures

Acknowledgments

This book required nearly a decade of research and writing, and that it exists at all is thanks to the enormous support that I received during these years. To begin with, I'd like to thank my colleagues in Screen Cultures at Northwestern University for their feedback and collegiality throughout the research for this project: Lakshmi Padmanabhan, Miriam Petty, Hamid Naficy, Ariel Rogers, Jeff Sconce, Lynn Spigel, Michael Turcios, Mimi White, and especially Jacob Smith, who read early drafts of this book and has been a vital mentor, sounding board, and friend. I feel fortunate to have enjoyed the support of department chairs here at Northwestern, including Thomas Bradshaw, Zayd Dohrn, and David Tolchinsky, as well as Dean Barbara O'Keefe in the early days of this project and Dean E. Patrick Johnson toward the end. There are too many other faculty members at Northwestern to thank them all, but I'd like to single out Masi Asare, Melissa Blanco, Laura Brueck, Tracy Davis, Dilip Gaonkar, Jim Hodge, Stephan Moore, Brett Neveu, Patrick Noonan, Dassia Posner, Alessia Ricciardi, Shayna Silverstein, and Elizabeth Son, all of whom gave me encouragement and supported me at key moments in recent years.

I also received a lot of support from my research assistants, as well as my (current and former) graduate students, including Kelly Coyne, Cara Dickason, Ilana Emmett, Martin Feld, Diana Funez, Dani Kissinger, Clare Ostroski, Golden Owens, Sarah Sachar, Amy Skjerseth, Jennifer Smart, and Elena Weber. Alex Knapp and Nicola McCafferty in particular were essential in conducting primary research and getting this manuscript ready for publication. Proofreader Ron Nowak helped me to prepare the manuscript for submission. Staff members Granville Bowerbank, Marysia Galent,

Katherine Lelek, Elizabeth Mathis, Shannon Pritchard, Dawn Washington, and Brad West made many aspects of my day-to-day work easier. I'd like to thank Russell Gillespie for his meticulous and creative work creating the mini digital audio workstation for chapter 2, as well as Brendan Baker for letting us use the sound stems of his episode for that feature. Pacific Content was generous in granting me a license to a graphic for chapter 1, thanks to Emily Shank and Dan Misener. DinahBird and Jean-Philippe Renoult, as well as Galen Joseph-Hunter and Wave Farm, were likewise incredibly kind to let me use images of their artwork for this book. The pitch tracking material was made possible by the generous funding of a National Endowment for the Humanities–sponsored Digital Humanities Advancement Grant, "Tools for Listening to Text-in-Performance," a 2018–19 project in which my inspiring collaborator, Marit MacArthur, and I drew on the talents of Lee Miller, Robert Ochshorn, and Mara Mills, among others. I am delighted that Mitra Kaboli agreed to lend the beautiful cover image.

This book is not based on formal interviews, but during the years in which I did my research, I was forever hanging around to pick the brains of producers, artists, and curators, asking, "What should I be listening to?" and listening to their thoughts about their work. Some of those generous enough to talk with me include Brendan Baker, Jesse Baker, Julia Barton, John Biewen, Amanda Dawn Christie, Elena Fernández Collins, Zachary Davis, John Dryden, Mischa Euceph, Anna Friz, Jeffrey Gardner, Fred Greenhalgh, Alan Hall, Bill Healy, Ann Heppermann, Ellen Horne, Galen Joseph-Hunter, Yvette Janine Jackson, Jeff Kolar, Andrew Leland, Ariana Martinez, Ele Matelan, Eleanor McDowall, Siobhan McHugh, Diarmuid McIntyre, Jonathan Mitchell, Eric Nuzum, Kara Oehler, Shima Oliaee, Kaitlin Prest, Adam Sachs, Joan Schuman, Jess Shane, Karen Werner, and Gregory Whitehead. Above all, I want to thank and praise Johanna Zorn and Sarah Geis, who were consistent advisors for many years. Both know far more about podcasting than I ever will.

Early versions of parts of this book appeared in *The Routledge Companion to Radio and Podcasting*, in the journals *Participations* and *RadioDoc Review*, and parts of this research that I decided to publish separately in *The Oxford Handbook of Radio and Podcasting*. I'd like to thank the reviewers of those publications for their many suggestions, as well as editors Andrew Bottomley, Siobhan McHugh, Michele Hilmes, Barbara Klinger, Mia Lindgren and Jason Loviglio. I have discussed the work that went into this book with a number of colleagues over the years, including Robert Boynton, Michael

Bull, Chris Cwynar, Tom Gunning, Henry Jenkins, Elena Razlogova, Alexander Russo, Josh Shepperd, Jennifer Stoever and Shawn VanCour. Several others whose work inspired me along the way include Lars Bernaerts, Richard Berry, Inés Casillas, Stacey Copeland, Tim Crook, Lance Dann, Susan Douglas, Kim Fox, Caroline Kita, Kate Lacey, Anne MacLennan, Virginia Madsen, Leslie McMurtry, Jarmila Mildorf, John Muse, Arionne Nettles, Aswin Punathambekar, Sonia Robles, Martin Spinelli, and Pim Verhulst. I'd also like to thank the many people with whom I worked on creating the 2023 Radio Preservation Task Force meeting at the Library of Congress, as well as the Kitchen Sisters and Rick Prelinger. A special thanks to Marianne Bower.

My experience at the University of Michigan Press has been wonderful from beginning to end. I can't thank them enough for their embrace of this project, their hard work putting it together, and their consistent professionalism. I am grateful to the staff, including Annie Carter and Danielle Coty-Fattal, for their help with assembling files and marketing the book and its online version. I would also like to thank Anne Taylor for diligent copy edits, Lisa DeBoer for producing an expertly crafted index, and Mary Hashman for her hard work as production editor. My editor, Sara Jo Cohen, has been a true advocate for me and my work for many years. I am deeply in debt to her for her patience, enthusiasm, and extraordinary ability to put in place a process that made this project all it could be. I am grateful to my anonymous readers, whose reports helped to bring everything home.

This was a very difficult book to write. I believe in doing things the hard way, but the pandemic, tenure requirements, illness, and deaths in the family made the period of composition for this book particularly mentally and physically challenging. It often seemed impossible, and I nearly abandoned it. My larger family—Delia, Ger, Sonia, Arun, Chloe, Evan, and my nieces and nephew—were supportive from near and far. My wife, Maureen, and my daughters, Maggie and Lulu, sacrificed a lot to help keep me going, and they believed in me far more than I believed in myself. Their strong commitment kept the project alive, and I am profoundly thankful for their love through every challenge. I dedicate this book to them, as well as to my parents, Caryl Verma and the late Dr. Harish Verma.

Introduction

What Was Podcasting?

"Podcasts Are Always the Next Big Thing" declared the headline of a 2020 article for *New York* magazine's *Vulture*, in which critic Becca James cheekily annotated some twenty-five news pieces about podcasting that had appeared online since the neologism first became widely used after a much-cited *Guardian* article in 2004 that is often credited with coining the term.[1] With 2020 hindsight, the articles in the list made for easy sport. In a series of withering glosses, James explored fifteen years of commentary about the coming "revolution" from the *New York Times*, *Rolling Stone*, *Vanity Fair*, *The Atlantic*, the *Washington Post*, and other media outlets that covered every "boom, bust and blip" in the sector with breathless prose, pegging predictions of inexorable change to flash-in-the-pan shows and couching developments in the terminology of utopian cultural mythmaking ("Golden Age," "Renaissance") mingled with the argot of natural resource extraction ("Peak," "Gold Rush"), at times even deploying cringeworthy colonialist rhetoric portraying podcasters as "pioneers" in a hitherto uninhabited "Wild West."

Heralding the "true arrival of the medium" using the same swaggering language year after year—here is James's joke—the cited articles cannot help but seem ridiculous when assembled. In a 2006 review of pods, for instance, *USA Today* considered a mere nine thousand episode downloads to be "big numbers"; a 2011 *New York Times* profile of comedy podcaster Marc Maron (one of many figures, from RSS pioneer Dave Winer to *This American Life* host Ira Glass, described as "The Podfather" over the years) made casual mention of gameshow-host-turned-podcaster Joe

Rogan, who the *Times* would a few years later declare to be the center of the "New Mainstream Media." James's list shows the rise of the genre of the pod recommendation "listicle," with publications releasing top-ten lists that had enduring zombie lives in search engine results, and reveals the extent to which minor celebrities and comedians boosted early networks like Earwolf, only to be eclipsed in 2014, when *This American Life*'s spinoff *Serial* was released, seeming to create from whole cloth a form of radio that would be like Netflix, at that time still new in the business of creating original programs. Soon writers were sagely asserting that podcasts were the new "new journalism," the new "personal essay," or the new "Gutenberg revolution." Rounding up her chronology in 2020, which James calls podcasting's "Sweet Sixteen" birthday, *Vulture* itself announced it would be doubling its podcast coverage, thus marking, as the author put it in mordant self-directed sarcasm, "the true arrival" of the medium.

Showing podcast discourse to be a thing of its time, the 2020 article has in turn become a thing of its time, one that I use to establish the first premise of this book—that podcasting is no longer really a "next big thing," one whose critical engagement ought to be oriented toward engaging with a swiftly impending future, but instead a "last big thing," one that is still growing and evolving as part of a complex media ecosystem, and the social understandings by which it is named and experienced, yet no longer plausibly resembles the Renaissance–Gold Rush–Wild West it was long advertised to be. If the stories that the *Vulture* article compiles seem juvenile, then it is because podcasting has outgrown them. It may be time for metaphors about entering a Baroque period, a Robber Baron era, or the part of the Western where the train line finally arrives to the dusty town. The proleptic imaginary that drove early discourse is no longer convincing, let alone an informative way into the critical task of discerning aesthetic passions of the medium. In the chapters that follow, I unpack three such passions that stretch across several hundred narrative podcasts that circulated in the post-*Serial* and pre-COVID period, roughly from late 2014 to early 2020. For the time being, I identify these passions as innovations in how audio approached affect, knowledge, and memory. One thing the passions share is that each was shaped by a changing listening and production culture and the cloud of boom-era rhetoric surrounding it; how podcasts explored feeling, depicted thought, and considered memory can only be grasped in the context of a medium that understood itself to be explosively emergent and brashly ungoverned. Indeed, the common belief that this was a period of emergence is on some level the "reason" these three passions came to be.

To anthropomorphize the medium for a moment, it would be fair to say that this book chronicles how narrative modes helped a new medium work out its youthful anxieties at a certain juncture in its ongoing identity crisis.

Even as the rhetoric of emergence and its concomitant proleptic imaginary can explain what themes, ideas, and agendas rose to the surface, that way of thinking can at the same time be an impediment to the analytical project at hand. My attempt at a solution is to write a book about the past. I am trying to see how a medium used narrative to deal with a sense of its own novelty in podcasts that thought of themselves as new, but I am paradoxically writing from a position that considers these same pieces to be old. I want to argue that podcasting is a phenomenon for which the time has come to ask not only the classic Bazinian question "What is it?" but also "What *was* it?" or more precisely "What was it back when many people were still captivated by asking 'what is it?'"[2] With this objective, this book takes an unusual approach. The vigorous argument about what is truly "new" about podcasting, or about what the medium "is" or "means," is a backdrop to all the podcasts I discuss, and I go over arguments about it in some detail, yet I want to distance this book from making claims that *answer* these questions, as I am skeptical that clear answers are available. What matters for this project is less "solving the podcast question" and more tracing how the narratives that I am approaching, in their unique historical milieu, took this question as a condition of possibility and used narrative forms to think it through, often in ways that were messy, jagged, and unfinished.

I believe this approach—one that considers narrative podcasts to no longer be new but uses them to approach podcasting's age of next-big-thingness as an object meriting historical investigation in its own right—can unbind podcast studies from the emphasis on prospective thinking that it has inherited from the popular hype surrounding the medium, while also unburdening the field from an exaggerated sense of duty to define and patrol the boundary of "the podcast" as a rigorously knowable form waiting to be intellectually claimed by one field or another. With all this in mind, the remainder of this introduction sets up the book's explorations, first by pursuing an argument for letting go of the proleptic imaginary captured in James's article, in favor of a retrospective approach to podcast analysis, one that could only be written now because the period it talks about is over. After making this case, I move into a more detailed picture of the "narrative era" that I propose to investigate in this book, as it sounds today, receding from earshot.

Inventing a Proleptic Imaginary

What drove podcast culture—from creators and fans to critics and scholars—to embrace a proleptic imaginary focused on next-big-thingness in the late 2010s? For an answer, we can turn to the many source articles that the Becca James piece mocked in *Vulture*. Digging through them, it soon becomes clear that while most mainstream writing about podcasts started out as an amalgam of entertainment news, tech rumor, journalism self-reflection, and lifestyle coverage, by 2017 business growth had become the central focus, with articles focused on "stupid" amounts of capital pouring into the "hot" sector. It may be no coincidence that 2017 was the year that Apple began to track podcast analytics as part of the rollout of its iOS 11 operating system, thereby generating the numbers that gave strength to the case for monetization of a medium that had hitherto, for the most part, been a form of time-shifted radio, amateur talk show production, and celebrity self-promotion.[3] Few articles in the *Vulture* litany fail to cite bullish indicators as charted by Edison Research, Zenith Optimedia, and other data and research firms looking at the emerging field in a detailed way: the rise in advertising revenue (some $35 million in 2015, $220 million by 2017, and nearly $1 billion in 2021); the rise in listenership (from around a third of Americans in 2015 to half in 2019); and the rise in the sheer number of podcasts (around 250,000 in 2014, rising to between 2.5 and 3 million at the time of writing).[4] Podcasts grew up at the same time as new tech-based financial instruments, like non-fungible tokens and cryptocurrency, and it is no surprise that despite being an order of magnitude less capitalized, they were spoken of in the same way.

The boom narrative, driven by reports making the case for podcasting as a business, often gave short shrift to counterevidence. For instance, by 2020, according to the Pew Research Center, although 24 percent of respondents said they got some news from podcasts, only 6 percent reported doing so often—hardly a revolution of Gutenberg proportions.[5] The growth of podcasting at the expense of radio was similarly not as rapid as it seemed.[6] In 2016, the middle of the podcast "revolution," Nielsen found that radio remained the most widely used electronic medium in the United States, with some 93 percent of respondents tuning in to radios each week.[7] Commercial radio's overall revenue was some $21.9 billion in 2021, while podcasting earned a paltry $1 billion according to a PwC report, with only modest increases projected for podcasts on the medium term.[8] Cloaking podcasting in hype was also a political choice, particularly as most programming was at the time oriented toward middle-class Whites, whose overrepresentation in podcast listenership did not begin to gradually change until the early 2020s. The dismissal of radio also excludes language communities. As Dolores Inés Casillas has pointed out, while Anglophone radio has declined markedly in recent decades, Spanish language on-air radio programming expanded from just sixty-seven Spanish-oriented stations in 1980 to nearly a thousand broadcasting in the language by the dawn of the 2010s.[9] These points rarely found their way into bubbling podcast business coverage. Instead, a mystique of next-big-thingness was generated by a focus on growth statistics, along with spotlight coverage of talent firms entering the sector, conferences like Podcast Movement, profiles of leading figures from Sarah Koenig to Conan O'Brien, and cult podcasts like *Welcome to Night Vale* and *Last Podcast on the Left* that went on stage tours. Together these elements offered material for strong storytelling about the medium's prospects, even if that narrative rested on mixed or mis-contextualized financial projections and grandiose hypotheses about social impacts, most of which were destined to come up short.

Still, it would be overdoing it to say that *all* podcast coverage of the time deserves the kind of drubbing that *Vulture*'s satirical article metes out. Indeed, James poaches in only one sector of a thicket of discourse that surrounded the medium, failing to draw from more balanced research and writing around podcasting in Edison Research's own meticulous reports and blogs; white papers from Pew and Nielsen; trade press sources like *Current*; online sources like *The Timbre*, *Hot Pod*, and Nieman Journalism Lab; and many academic journals, from the *Journal of Radio and Audio Media* and the *Radio Journal* to *RadioDoc Review*, that flourished in these years. All by itself, the Bello Collective—a group of writers focusing on incisive podcast writing, including Galen Beebe, Wil Williams, Dana Gerber-Margie, and Elena Fernández Collins, among others—published thousands of balanced podcast reviews, recommendations, interviews, and industry analysis pieces from 2016 to2022, mapping out the terrain around them at the same time.[10] In 2020 Beebe cited some fourteen searchable publications with regular detailed podcast critical writing; the subsequent year, Bello writer Erik Jones cited ten venues for industry news, five for podcast culture coverage, nine for creators, and twenty for review content.[11]

And, after all, when it comes to media, this hype is hardly a surprise. Anyone with a passing sense of popular discourse surrounding media technology in the past two centuries will recognize this proleptic language, signifying that a medium is being rhetorically framed as emergent in nature, an ideological characterization that benefits investors and enterprising creative workers alike, usually abetted by mainstream press, where sanguine forecasts always sell more papers and earn more clicks than humble retractions of prior predictions, and also by scholars who are drawn to emergence, often to claim or defend turf.[12] These moments naturally prompt a zeal to debunk the dramatic claims that surround them, but they also afford us opportunities to study how claims frame contests and negotiations among social groups, making space for forms that emerge and diminish as new media become, inevitably, old. "New media are," as Lisa Gitelman has put it, "less points of epistemic rupture than they are socially embedded sites of the ongoing negotiation of meaning as such."[13] One marker of that negotiation in this rhetorical milieu is the emergence of a kind of temporality, a key formal attribute of the prevalent 2010s' discourse about podcast "new worlds" and "revolutions." By adopting an approach to analysis framed in the present simple (here is this new podcast phenomenon) but whose gravity weights on the future tense (it signals a shift just over the horizon), writers lean their prose toward a coming promised era and, in so doing, pre-structure a set of questions and expectations and the means by which they could be met. In this way, next-big-thingness had a way of drawing on the fabric of a business growth case to swaddle podcast culture, as well as tools devised for understanding it, from head to foot in a future-oriented thought process, a proleptic imaginary that felt gripping in the moment but could not help but fall on its face in time.

This book starts as an effort to unwrap podcast analysis from that imaginary, in the belief that it no longer works effectively as an analytical point of departure. What was remarkable about podcasting used to be what seemed so new; what is remarkable about it now is what seems to be old. In this spirit, while many of the readings in this book started out as blog posts, articles, and chapters that were "hot takes" in the 2010s, this book transforms them into cool takes in an effort to move past questions formed in the context of frenzied growth, which had urgency a few years ago (affecting the author, too), and to get into thematic and narratological ones that in retrospect we can now see characterized the life of a medium that was not at the time quite certain of its own instruments, objects, potential, or purpose.[14] My aim is to look at podcasts of the late 2010s and explain their common fantasies and aesthetic agendas, some of which can only really be captured by letting go of the preference for prospective temporality in podcast studies and instead deliberately scaffolding the material within a retrospective mood.

I am not alone in seeking an alternative temporal framework with regards to this material. Even within the period I am discussing, there were many for whom podcasting was already old, its histories multiple and conflicting. As early as 2015, podcaster Benjamen Walker released an episode of his *Theory of Everything* podcast on exactly this theme, noting that the term was more than a decade old for him and that the "secret history" of podcasting was in fact a triple history of a business model, a technology, and an art form, each of which had its own *Rashomon*-like perspective on the key events. And now that we have distance from the proleptic imaginary that *Vulture* lampoons, many news stories are beginning to question the expansive "revolution" narrative by citing symptoms of a surprising homeostasis in the industry. In 2022, for instance, Bloomberg reported that despite the efforts of today's big players, the top ten podcasts on the Apple and other charts were on average seven years old and that only a few of the top fifty were less than two years old.[15] Another survey study found that a majority of successful podcasters (defined as those making more than $50,000 a year) already had more than two hundred episodes to their name, suggesting less room for newcomers to enter the mainstream of the field and achieve subsistence, let alone limitless success, as niche marketing strategies took over.[16] Even many of the companies that made a name for themselves during the boom have by now been long absorbed into larger entities: Gimlet Media financed with $6 million in 2014 and sold to Spotify, along with The Ringer and The Anchor, for a total of some $800 million in 2018; *Serial* founded in 2014 and sold to the *New York Times* for $25 million in 2018; Wondery founded in 2016 and sold to Amazon in 2021 for $300 million; and Stitcher, a key 2010s podcatcher, distributor, and producer, bought by Sirius XM radio in 2020 for $265 million, and dissolved entirely by the summer of 2023.[17]

The pandemic also cooled the sector somewhat. A recent report showed slowing growth both in podcast audience and advertising revenue.[18] One commercial analysis showed that while a million podcasts were launched in 2020, the next year that number declined to fewer than eight hundred thousand.[19] By 2022 and 2023, prominent critics and producers, such as Nicholas Quah of *Vulture* and Julie Snyder of *Serial*, were openly wondering if podcasting was going the way of radio—where all the energy would move into personality-driven talk shows and celebrity vehicles—and even Spotify, once described as a "one-company podcast bubble," had slowed its pace of investment.[20] The week this manuscript was completed in June of 2023, Spotify dissolved Gimlet entirely, laying off some two hundred workers. Some of the signature podcasts of the 2010s were canceled that summer, including multi-award winners such as *The Truth* and *Invisibilia*, and apps that had played an important role in growing the sector had disappeared. While these theories of historical multiplicity, industry homeostasis, or even creative decline should not lead to the facile conclusion that podcasting "is dead" (that would merely trade one fallacy for another), some healthy skepticism toward the proleptic imaginary has begun to settle in. This development is on balance a salutary corrective, suggesting there is merit to a focus on what podcasting was at a moment in the recent past rather than rushing headlong into speculations about what it may become next.

The Burdens of Definition

My emphasis on retrospective mood also differentiates my approach from that of other writers on the medium, who have for the most part tended either to emphasize continuity between podcasting and radio forms of the *longue durée* past or to posit distinct breakage with those forms, a nuanced and evolving debate that also happened, and for many of the same reasons, among scholars of cinema and television during the 2010s.[21] Podcast studies, books, and essays published during this period were often freighted by three onerous questions: What is the definition of podcasting? When did it begin? Does it extend, replace, or form a companion to radio? This triad of definitive, originative, and differentiative presents three faces of the same issue, and thus committing to a position on one flowed into a position on the others.

For those authors for whom web-based distribution via an RSS feed is a sine qua non of counting as a defined "podcast," the history tends to foreground developer Dave Winer, MTV star Adam Curry, and journalist Christopher Lydon, whose *Open Source* podcast in 2003 was by most accounts the first to use this system.[22] A narrative foregrounding this lineage tends to focus on issues such as podcatcher uptake in smart devices, the role of Apple through both its iPod devices and its corporate choices about using the model, and the rise of audience measurement, as well as any other matter surrounding the conditions that allow a growing industry to thrive, particularly monetization. A strong typical example might be a 2015 article for the Tow Center and the *Columbia Journalism Review*, in which researcher Vanessa Quirk quite early on saw three main revenue generation "philosophies" emerging, one based on advertising, another on free distribution along with premiums for patrons, and a third that used podcasting as a non-revenue-generating way to enhance brands.[23] This definition has also been key to preservation efforts. From 2015 to 2019, scholars Jeremy Morris and Eric Hoyt used RSS feeds to create PodcastRE, a digital archive of podcasts of over a million files from seventy-five hundred feeds, creating a searchable database that provides metadata surrounding these files, as well as measurement of the frequency of terms used in the corpus.[24] But remove the RSS feed as a defining element of the medium, and the history changes, pointing back to the audioblogging and web-based radio of the 1990s that Andrew Bottomley has brilliantly chronicled, a narrative in which the convergence of digital radio and the internet is more crucial than any particular distribution subsystem dedicated to managing that convergence.[25] This focus has the benefit of driving the conversation down to the grassroots, to amateur creators and college radio stations, improvisers and low-tech media creation. It also provides a more dynamic picture of events, showing how the internet drastically affected how radio was made at the same time as it created a new medium.[26]

Focus on an archaeology of the very word "podcast," and the story changes in a different way. In an early effort at defining podcasting, for example, Jonathan Sterne, Jeremy Morris, Michael Baker, and Ariana Freire explored new actors and technologies appearing around the term, but also focused on the failed promise of the parent term "broadcasting" as a point-to-mass model that became a much narrower cultural mode in the network radio era than its original concept suggested, and considered the podcast as a possible redemption of the democratizing ambitions that early broadcasting had promised.[27] For authors who focused on podcasting as an idea in a longer history of radio, such as Michele Hilmes, "podcasting" represents yet more proof that radio continues to be a heterogeneous and constantly shifting medium.[28] The "radio approach" helps knit together obvious thematic continuities from one form to another, allows for a discourse of media convergence and remediation (two concepts crucial to new media theory approaches to podcasting, following the well-known work of Henry Jenkins on the one hand and Jay Bolter and Richard Grusin on the other), and rightly emphasizes that by opposing "radio" to "podcasting" we run the risk of oversimplifying both.[29] So, when writers such as Lance Dann and Martin Spinelli put forward perhaps the most successful proposal on what made podcasting worthy of its own critical language in 2019, citing some eleven features of podcasting that really do show that its production and consumption no longer resemble those of the form of "radio" with which most people in Western societies are familiar (e.g., BBC, NPR), a response could be that "radio" has always been a more capacious term than simple enumerations of the features of its dominant or emergent modes of circulation suggest.[30] For radio historians, at stake in defending the category of "radio" is not just a professional investment—I am reminded of D. N. Rodowick's observation that film studies earned acceptance in the academy around the time when its main object disappeared—but also the insistence on a relation to history and to the forms, precedents, and communities that persist into the contemporary moment of creative audio.[31]

What I want to emphasize about this debate is how the triad of definitive, originative, and differentiative tended to lead to one another, generating a shifting, unsettled conversation. The discourse was in these years caught in a triangle turned by a feeling of urgency that affected all positions on it. Both those who employed longer historical frames to manage that urgency and those who emphasized shorter ones were reflecting a media culture undergoing an identity crisis before their very ears, a crisis of self-understanding characterized by a destabilized present and uncertain futurity that presented as a need for decisive historical anchorage, which scholars quite understandably felt called to provide. My purpose here is not to referee this debate but to show how its very existence is symptomatic of the proleptic feeling of next-big-thingness that characterized podcasting culture in the late 2010s, a feeling that forms the backdrop to much of the audio produced in this period. It would be a challenge to try to understand narrative audio in this period without first recognizing that it was a form whose structure, meaning, and even ontology were constantly being debated and that scholars both reflected and took part in that dispute. In this book, my goal is to hear how that debate echoes inside podcasts of the period while also arguing that it is a debate whose burden it has now come time to shed.

That said, to avoid unnecessary vagueness, I want to set out for the reader my thinking about how I use the term "podcast" at least in the context of this book. For one thing, like Richard Berry, I feel that pinning podcasting's definition, origin, or differentiation to specific technologies is a strategy that is destined to fail.[32] The *Oxford English Dictionary*'s definition of "podcast," a Word of the Year in 2005, describes it, as of 2022, as "a digital audio file of speech, music, broadcast material, etc., made available on the internet for downloading to a computer or portable media player; a series of such files, new instalments of which can be received by subscribers automatically."[33] This definition no longer uses the term "radio," as it once did, and is both broader than earlier versions (incorporating music, for instance, and perhaps audiobooks, which fit the definition) and narrower than others (linking the phenomenon to downloading and to a dedicated device, chiefly the Apple iPod, which gave the phenomenon its prefix). As several authors have noted, a few cobwebs have descended across this definition, not least because at this point more users stream podcasts rather than bothering to download them, while the iPod and similar devices have been discontinued with the rise of streaming by cell phone. In truth, this approach was always a little unstable. In 2016, in its annual "Infinite Dial" presentation and report, Edison Research noted that the technology was becoming invisible to listeners, with many fans simply clicking a media link to access podcasts rather than taking time to subscribe or download.[34] Newcomers to the medium today, according to surveys, tend to stream podcasts on platforms initially designed for music or video, like YouTube or Spotify.[35] The wall separating podcasts from other media is also becoming porous. Many "legacy" radio organizations, from iHeartMedia to BBC Sounds, produce podcasts—NPR has done so for a dozen years—as do large transmedia brands like Marvel.

Ten years ago, students in my classes had a clear idea of the difference between a "radio show" and a "podcast," but now they use these words interchangeably, since podcasts often appear on air and terrestrial radio is accessed on smartphones. There is actually an important intuition behind this media confusion. Radio, smartphones, and Wi-Fi are all reliant on "wireless" communication (a term coined to describe radio back in the age of telegraphy) using segments of the electromagnetic spectrum overseen by the same regulatory authorities as radio bands. In the United States, whether an audio narrative comes to you by the airwaves, a cell phone tower, or broadband, the infrastructure making that possible is overseen by the Federal Communications Commission (FCC) and other agencies, and is shaped by a regulatory environment formed by the Communications Act of 1934, the Public Broadcasting Act of 1967, and the Telecommunications Act of 1996, as well as other legislation and FCC guidelines.

With this in mind, like media scholar Tiziano Bonini, I have come to think of podcasting as a hybrid audio form that involves an evolving network of actors, ranging from producers and listeners to publishers, platforms, technologies, and material infrastructures, all in competition with one another.[36] This approach strikes me as a necessary concession to the fact that the term itself has shifted untidily even in the short time that advocates of podcast studies have been putting down roots. Indeed, it is possible that the period in which the definition of the term "podcast" had maximum potential for sharpness has already risen and declined; today there may be too many digital audio artifacts (individual and serialized, amateur and professional, commercial and public, national and international) that are marked as podcasts by users (be they on embedded players on websites, in broadcast schedules, in proprietary apps, accumulating in podcatcher feeds, tucked away in cloud-based storage, or even at the Library of Congress, which began preserving podcasts in 2020) to set forth any future-proof conclusions about their shared characteristics, beyond something really simple like referring to podcasting as digital on-demand audio.

At any rate, shepherding this ever-growing herd into a single model is not the errand of this book, which does not hope to answer the question of "what podcasting is" in any ultimate sense but rather attempts to observe what formal strategies and thematic preoccupations guided podcasts during the years in which the question itself was at the zenith of its urgency. In what follows I am primarily after the *idea of podcasting* that a given piece of soundwork explores rather than trying to propose and defend a rubric characterized by a set of criteria to which the piece in question may or may not conform. A digital audio file is a podcast only when we enchain it to a set of sensibilities that surround that term, a set of sensibilities that are themselves evolving. And because a commonsense semantic meaning of the term "podcast" was neither agreed to by everyone involved nor stable over this period, for the purposes of this book, I thus consider something to be a podcast if it is made critically understandable and explorable using that term, as it was understood by creators, critics, and/or listeners at the time of the piece's initial or ongoing circulation.

Methods for Retrospective Podcast Theory

This book represents an effort to train an analytical ear on a time in media history that has only just recently become media history. I am going to explore around three hundred podcasts and several thousand episodes that circulated from the fall of 2014, when *Serial* became a hit, to the start of the COVID pandemic in early 2020. My examples reflect many of the most popular and awarded narrative podcasts from this period, but they also include lesser-known pieces that caught my ear. I have endeavored to make a sample set that is more inclusive than the industry and its listenership may truly have been in this period, and this book includes works by and about women, members of queer communities, podcasters of color, people of differing ability, and Indigenous communities, many of whom brought to the medium, as scholars such as Kim Fox and Sarah Florini have shown, rich ways of making and thinking that White, cis, abled, male creators did not.[37] My examples focus mostly on Anglophone podcasts by necessity, most are American in origin (the United States dominated the global sector for narrative podcasts in these years, though that is changing today), and they follow an idiosyncratic set of interests. Scholars invariably follow up most diligently on primary materials that fascinate and vex them, often for reasons they cannot quite identify, but are in some ways always tacitly shaped by their own identities and intellectual commitments.

This book is also shaped by methods and preoccupations that also guided my previous book, *Theater of the Mind*, which dealt with several thousand radio plays from the 1930s to the 1950s. Like that book, what follows frequently looks at very small details in editing, sound, and writing (what I call close listening) in an intentionally obsessive way that feels like textual analysis, drawing on classic and new theories of audionarratology, as well as pursuing structural similarities across a great many podcasts (what I call distant listening), reflecting my confessed tendency to think in terms of speculative taxonomy.[38] While this approach arises, frankly, from what I feel represent my modest aptitudes as a scholar, over the course of the book it should become apparent that obsessive listening and structural resonances also help to capture some of my suggestions about the way criticism itself can be bound up in the same conceptual projects as the podcasting with which it engages.

Inevitably, a book of this kind may give the impression of naming the "greatest hits" of the period. That is not my intention. For one thing, assessments of merit are not my expertise. There are several other writers who know better than I do how to use key examples to respond to the prompt of how to make a good story podcast.[39] Indeed, the entire Transom website is a trove of these reflections for those interested in nonfiction, as is the *Radio Drama Revival* podcast for those interested in fiction.[40] My role is to reckon how podcasts think about feeling, memory, and knowledge itself rather than to assess their success with audiences or taste communities, and so what follows is relatively agnostic about the issue of artistic merit. Moreover, I am not trying to box out other narratives about what other scholars may feel are more important themes and forms from the period. In fact, I hope to prompt alternative ideas that I may have missed, given my limited ability and positionality as a Punjabi-White cis heterosexual middle-aged man, and to empower better and more fruitful readings of shows that I may have unjustly denuded of important details. For a listening-based critical practice like mine, podcasts of the late 2010s are so superabundant, so dense, and so lengthy that they represent an impossible corpus for which to do a credible "overview" with ears alone. In order to turn podcasting itself, as a totality, into a research object, it is necessary to turn it into a dataset rather than a playlist. For that, the PodcastRE project far and away stands the best chance of yielding understanding, and Hoyt et al.'s recent analyses have done just that, lending insights on podcast durations across genre and network, as well as a number of other facets of podcast norms.[41] In the future, this corpus may be unpacked primarily with machine listening, which will (as any research approach) reconstitute the object of its inquiry on its own terms. For now, from a humanist perspective, the best I can do is trace what I believe to be a set of interlocking concepts in this corpus based on what I hear and to connect them to issues facing the medium's historical context, cultures, and communities.

Those communities form a part of this book's methodology, too. Many of my ideas about podcasting were generated while I attended meetings of podcasters in North America and Europe, and others developed in informal conversations with established and new creators over the better part of a decade. While this book is not a sociological portrait of this itinerant, evolving community (there is new fascinating work along these lines), these were important grounding experiences for me in this project.[42] The participant-observation mode in these pages is also meant to at least partly fill in the "mid-level" between my close readings and discussions of larger structural trends, reflecting a recognition that podcast culture did not merely exist in the audio itself but in a variety of in-person and digital spaces, from newsletters, twitter threads, and livestreams to sound installations, live performances, and conference programs. Podcasting debated its future in many places, from Brooklyn apartments to journalism programs in Australia, from Chicago's Third Coast International Audio Festival to the HearSay Audio Arts Festival in Ireland. I could be part of some of those, but even from afar it was clear that the complex dance between the terms "podcast" and "radio" was happening all over the world, from In the Dark's "The Listening Body" popup at The Barbican in London curated by Eleanor McDowall and Nina Garthwaite, to the Brooklyn Academy of Music's annual RadioLoveFest, which steadily featured more podcasts than radio shows, and the European Broadcasting Union's International Feature Conference, a fifty-year-old documentary awards organization that in its 2022 meeting in Ireland chose "The End of Radio" as its main theme, as well as similar conferences in Asia and elsewhere. Following these communities allowed me to learn about the evolution of purpose-built digital audio workstation software like Hindenburg and Descript, but also about grassroots preservation efforts such as the *Preserve This Podcast* zine, a guide that was distributed at conferences to help podcasters think about their archives. And the communities that coalesced at these events also formed online to change the way the industry worked, including interventions such as the efforts of Renay Richardson and Broccoli Content to recruit audio companies to the Equality in Audio Pact in 2020, an initiative to ameliorate inequities in the industry by committing to paying interns and hiring LGBTQIA+ identifying producers, among other reforms.[43]

Finally, this book also uses a number of tools drawn from digital humanities projects that are relatively new in critical audio studies, from pitch-tracking to spectrographic analysis, that I feel offer apposite ways of doing close listening to fragments of audio narratives, implicitly making the case for the importance of using digital analytical tools to engage digital primary materials, which are always already remediated code. Sometimes the things I hear happening in these plays can be made clearer by looking at the digital workstations in which they were mixed or by transforming them post-mix into visualizations using digital tools. As these chapters proceed, I introduce these tools for specific purposes, as a way of giving shape to critical listening practices. What's more, as the conclusion to chapter 2 argues, the use of digital tools reminds us of the inherent multiplicity of podcast audio as a digital object, something that itself was a matter of critical significance in dealing with works of this period. In engaging in this way, this book experiments with ways in which audionarratology can be reimagined in a digital setting, efforts that I hope others will find useful, build on, and eventually outstrip.

The Rise of Narrative Podcasts

One reason for putting a frame around the period from *Serial* to COVID—late 2014 to early 2020—is already apparent, as these years represent a flourishing of the proleptic imaginary surrounding the medium. Those who evangelized, created, or even just casually followed podcasts in these years tended to do so with the problem of "what a podcast even is" as a forward part of their thought process and professional ambitions. Second, during this period, podcast growth was particularly associated with *narrative* or "story-driven" works. The narrative dimension brought the media form into prominence, and story podcasts dominated many news items about the emerging medium. It is thanks to their narrative form, for instance, that many podcasts of the period either inspired or became TV shows (*Lore*, *The Case Against Adnan Syed*, *Dirty John*, *Homecoming*) and books (*The Infinite Noise*, *Lost in Larrimah*, *Welcome to Night Vale*).

Although the term "narrative" is used frequently in soundwork communities, its value is often elusive and its meaning deferred. At one point in Jessica Abel's *Out on the Wire*, *This American Life* (*TAL*) producer Ira Glass disparages "narrative" as a "boring" way of saying "story," but in subsequent pages he uses the term as a technical necessity, for much the same reason narratologists have for a century, as an instrument to make expressible interactions between form and content across a text and language.[44] Some texts in radio production lean on "ingredient list" approaches to what a good story possesses. In *Sound Reporting*, for example, a text by NPR training head Jonathan Kern, a good audio story is said to possess the same attributes as a fairy tale: characters; setting; a clear beginning, middle, and end; tension; and resolution.[45] In a 2014 webinar entitled "Power Your Podcast with Storytelling," *TAL* and *Marketplace* producer Alex Blumberg, then freshly minted as the CEO of Gimlet Media, summarized a practical sense of that program's idea of narrative, what he called the "nuts and bolts" or "physics" of a story, which typically consists of four elements: a sequence of actions, telling details, a point or punchline, and a moment of reflection.[46] Time and again in his webinar, Blumberg breaks down this "physics" using stories from *This American Life*, a show whose influence on the very concept of what makes a "good" narrative is hard to overstate. At one point I was calling this book project "Radio after *Life*" as a recognition of the degree to which that show's aesthetic reverberated across so many podcasts in the 2010s. *TAL* was at this time by many measures the most widely distributed podcast in the world, in addition to being a widely syndicated radio show, and it is a crucial if forgotten matter of its history that *Serial* debuted on *TAL*'s radio show at the same time it was released as a podcast. Like Blumberg, Julie Snyder, and many other key producers of the mid-2010s, *Serial* host Sarah Koenig spent an important part of her pre–podcast career making *TAL* stories. At the time her podcast debuted, she was listed in the production credits of some thirty-eight episodes in the 2000s.

But the "physics" model was not the only idea of narrative out there. Among producers, there was a less specific but perhaps more useful common linking convention for the present purpose: it is the presence of "scenes" in the narrative, usually taking the form of tape from outside the studio setting, that really differentiates a narrative (fiction and nonfiction) from most other varieties of chat podcasts or talk-forward audio.[47] In audio fiction, with some exceptions, it has long been the sonic illustration of fictional scenes, in which characters perform actions, that differentiates a drama from the act of reading a fictional text aloud. In nonfiction, the presence of scenes similarly differentiates narrative by instituting an inter-supplementary system whereby host-spoken commentaries are broken up, distributed, and supported. Scenes corrugate a written and vocalized script with audio offered as if it came from a second (or third or fourth) time-space that is separate from the place of narrative enunciation. This is a definition that influences my approach here, a feature that narrative theorist Marie-Laure Ryan has recently pointed to, a feeling that we are toggling between seemingly direct and narrated presentation, what she calls an "interplay between mimetic and diegetic narration."[48] For producer Alix Spiegel, different ways of staging that interplay could give programs some of their signature feel; the longtime public radio producer described *TAL* as a show in which a host "hands off" to tape, whereas a show like *Radiolab* let tape erupt, disrupt, and poke through its narration, so there was often no clear differentiation between the time-space of narrating and the time-space of reporting.[49]

Perhaps the best set of illustrations of this form of narration can be found in the archive of the Third Coast International Audio Festival, a Chicago-based competition and meeting place founded by Johanna Zorn and Julie Shapiro that did more than perhaps any other institution to define and nurture narrative audio in the early twenty-first century in the United States.[50] The first class of its Richard H. Driehaus awards for audio stories in 2001 contains a dozen highly varied narratives in which narrators speak directly to us in scripted and nonscripted fashion, as they also intermittently take us into secondary, tertiary, and quaternary temporalities ranging from (across these pieces) tapes of the Vietnam War to interviews with survivors of a prom-night tornado, scenes from *The Sound of Music*, and gripping moments of a teenager confronting her father about his drug addiction. The content and sourcing of scenes can vary wildly, with archival tape, interview tape and other actualities, or even reenactments, and the narrative can veer into unpredictable places, but in each case the tessellation of scenes into a prepared bed of narration is a sign of bona fide narrativity and scriptedness. The moment a "scene" takes place, we are no longer just in a pseudo discussion with a speaker, as the recipient of their frontal address, but "inside a story," the experience of which is inextricable from a set of structuring codes developed for and through a mediating apparatus; we are enlisted as cocreator by means of the expectation that it is our task to mentally coordinate a series of times and spaces as they appear. It is not just a "person talking to us" any longer but a "narrative happening." Whatever their differences, on this level, there is a match between the way fiction and nonfiction identify themselves. All podcasts discussed below fit into either the *TAL* model or the broader scene-based idea of narrative; most fit in the center of the overlap of these models. So while podcasters often also tell stories, sometimes long-form stemwinders, in the context of talk-based podcasts—this happens often in unscripted interviews—by and large these fall outside of the perimeter of this particular project.

The final reason for my periodization is that it was in these years that a great many shows that self-identify as narrative (using the definitions above or related ones) began to migrate from "radio" to "podcasting" in a self-aware way. You can see this by the way awards organizations focused on story-driven work gradually began to recognize the form. The Third Coast Festival gave its first award to a podcast in 2011, establishing a whole category for serialized stories in 2019, while even longer-running awards organizations like the Prix Italia evolved, giving a special award for New Radio Formats to its first podcast in 2015. In her recent book, Siobhán McHugh has long lists of similar awards, some old and some new, that by 2022 were places where podcasts could earn peer recognition.[51] One-off winners who appear on these lists only hint at larger trends, but the results alone only tell part of the story. Luckily, there is one major awards organization that archives all its entrants in a research setting, the George Foster Peabody Awards, archived at the University of Georgia, which has been giving awards to radio and audio features since 1940 as the oldest organization granting awards for work in electronic media. Its archive of some seventy-eight thousand items from the history of television and radio contains a robust set of metadata about where podcasts came from and were distributed, thereby offering researchers a picture of a corpus of creative narrative audio whose creators at least *felt* could reach the level of prestige.[52] It is a kind of core sample of narrative audio in the period.

That sample helps clarify the environment of prestige audio in the late 2010s. For the purposes of this book, my research assistant Alex Knapp and I analyzed five years of metadata from the 1,263 audio-based submissions made to the Peabodys from 2014, the year in which the word "podcasting" was first explicitly used as part of a category title, to 2019, the last full year

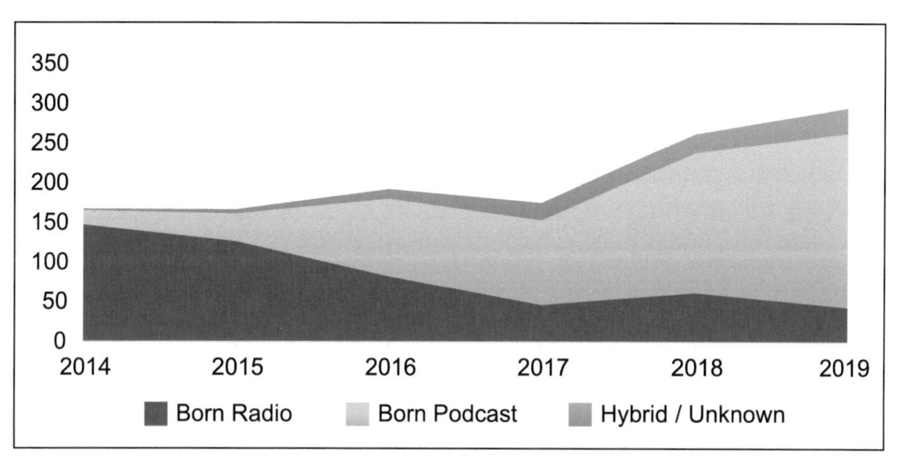

Figure 1. Number of Peabody Radio/Podcast Award entries, by media form. 2016 represents the dividing line after which a majority of entrants begin to classify their work as a podcast rather than a radio show. By the author with data compiled by Alex Knapp.

before the pandemic.[53] The data represented in figure 1 show that in 2014 the overwhelming number of entrants to the competition—88 percent—filed paperwork that identifies themselves as "born radio" pieces. Even if they were subsequently released on the internet or archived on a radio exchange, they were submitted to the awards in a way that indicates they were initially conceived for on-air transmission, often identifying an originating station, network, or syndicating entity. Just two years later, in 2016, the tide had turned, with some 51 percent of entries identified as "born podcast." And by 2019, the scales shifted decisively and perhaps permanently, with some 74 percent born podcast and just 15 percent born radio. The sheer number of entries had also risen enormously, from 168 entries in 2014 to 296 in 2019.

What were the programs behind these numbers? An examination of the archive entrants is an index of a restless, fast-moving industry, bursting from within an established one. Every year there are perennial submissions from celebrity-driven shows that emerged from talk and comedy programming that had been on podcasts since the early days, such as *The Q&A with Jeff Goldsmith*, the *Bill Simmons Podcast*, *WTF with Mark Maron*, and shows featuring pop culture figures such as Alec Baldwin and Lena Dunham. At the same time there are story-focused shows with roots in public radio distribution, like *This American Life*, *Radiolab*, *L.A. Theater Works*, *Reveal*, and *Snap Judgment*. Offshoots from the latter group emerged in "limited series" formats that were often poised to succeed. *Radiolab* released a rich variety of these, including *Radiolab Presents: G* (about the idea of intelligence), *More Perfect* (on the Supreme Court), and *Dolly Parton's America* (on the beloved cultural icon). Personal storytelling podcasts had a mini-boom among the entrants in 2014 and 2015, including shows that were already or have subsequently become institutions, such as *StoryCorps*, *Nocturne*, and *The Moth Radio Hour*, and those that petered out, such as *The Lapse* and *Everything Is Stories*. Other early and regular entrants either came from Earwolf and its parent, Midroll Media, like *Comedy Bang! Bang!*, or were offshoots of NPR programming, including *Freakonomics Radio*, *NPR Pop Culture Happy Hour*, and *Hidden Brain*.

The years 2015 and 2016 brought innovation in programs like *Invisibilia*, a show founded by celebrated producers Lulu Miller and Alix Spiegel that blended the formal features *Radiolab* and *This American Life*, as a growing slate of shows also revealed an industry seeking new listeners and formats, like *The New Yorker Radio Hour*, *2 Dope Queens*, and *Death, Sex & Money*. The latter came from WNYC Studios, which raised some $15 million to set up a slate of podcasts in 2015 that would eventually compare and compete with offerings emerging from the Public Radio Exchange's Radiotopia, a group of programs ranging from *99% Invisible* and *Ear Hustle* to *The Heart* and *The Memory Palace*. These programs as a collective made waves by raising funds with a legendary $600,000 Kickstarter fundraising campaign by producer Roman Mars in 2014 that soon led to a $1 million grant from the Knight Foundation for the group as a whole. Over time, newer start-up distributors crowded the competition. Shows produced or distributed by Gimlet (founded in 2014), like *Reply All* and *Heavyweight*, began to compete more frequently, with six submissions prior to 2017 and then an increasing number nearly every year afterward. The same is true for several other companies: Crooked Media (founded in 2017), a left-leaning political network famous for *Pod Save America* and its offshoots that shifted gradually from talk to story formats; Wondery (founded in 2016), which hosted several true crime franchises; Pushkin (founded in 2018), home to shows like *Revisionist History*; and Panoply (founded in 2015), a short-lived but popular network that pioneered branded audio content, funded early on by *Slate* magazine, which ran many franchises based in existing magazines and other media venues.

After *Serial*'s win, these early years also saw a sharp rise in the submission of crime podcasts, many of which came from news publishers, like the *Atlanta Journal-Constitution*'s *Breakdown* in 2014 and the *Milwaukee Journal Sentinel*'s *Unsolved* in 2015, as many others emerged from podcast-centric producers, like Radiotopia's *Criminal* and Gimlet's *Crimetown*, as well as from public radio organizations serving their communities, like WBEZ Chicago's *16 Shots*, which explored the police murder of Laquan McDonald. In 2017 the arrival of the *New York Times*'s *The Daily* signaled a deepening investment in audio by that brand (the paper had been creating podcasts for more than a decade already), like many other legacy companies and arts and cultural institutions that year, from ESPN, the *Wall Street Journal*, and Bloomberg to CNN, the BBC, and the Metropolitan Museum of Art. The archive also has a number of narrative-driven issue-based podcasts, on all manner of topics: *Bundyville*, about the renegade ranching family; Ronan Farrow's *Catch and Kill* podcast that was the center of the #MeToo movement; "The Weight of Dust" about 9/11 responders; and podcasts about the Grenfell Tower collapse, to take just a few examples. Several podcasts focused on horrific shootings of unarmed black and brown people, a longtime epidemic that rose to greater White media attention in the years between the 2012 killing of Trayvon Martin and the deaths of George Floyd and Breonna Taylor in 2020. Other social and political questions of the period are also represented in the archive, including inequality in urban education systems, the eviction and homelessness crises in California and elsewhere, current and historic redlining in homeownership designed to exclude people of color, the war in Afghanistan, environmental degradation, international adoption, the opioid crisis, no-knock warrants, and transgender identity. Historical anniversaries also served as prompts for podcasts, particularly when they had bearing on political and social questions, as these years saw podcasters revisit the politics of 1968, the Nixon era, the Stonewall riots, the moon landings, the Civil War, and World War I.

Soon came the rise of several fiction podcasts, a few of which (*Welcome to Night Vale*, *The Truth*, *The Thrilling Adventure Hour*, *Pleasuretown*) appear in the archive early in this period, though pivotal shows, like *Limetown* and *The Black Tapes*, which successfully blended the idea of the investigative podcast with that of fiction, are left out. Also omitted are important innovators, such as *The Message*, a No. 1 iTunes podcast that represented an experiment in content marketing from General Electric in partnership with Panoply, as well as key independent shows like *Our Fair City*, *Wolf 359*, and *The Bright Sessions*. In 2016 the genre began to have more presence in the archive, thanks to Night Vale Presents shows like *Alice Isn't Dead* and

Within the Wires; various children's shows created or distributed by Pinna and Gen-Z Media (whose *The Unexplainable Disappearance of Mars Patel* became the first podcast audio drama to actually win a Peabody) and other organizations; and podcasts from Gimlet's fiction arm, such as *Homecoming*, not to mention several intriguing independent shows, like Two-Up's *36 Questions*, which billed itself as the first podcast musical, and Issa Rae's *Fruit*, a queer-of-color sports thriller. In 2017 Amazon subsidiary Audible would invest some $5 million in the development of new audio fiction, but this would not present the competition with many releases until the 2020s. Across all its offerings, Audible's various audio production units (the differences are hard to untangle) directed a lot of its work to the awards, from high-concept professionally reported podcasts like *Damned Spot*, to adult marital advice content like *Where Should We Begin?* with psychologist Esther Perel, to dozens of children's shows. Overall, they submitted three podcasts in 2016, nine in 2017, ten in 2018, and sixteen by 2019. Outside of that ecosystem, these years were also the heyday of shows like *Song Exploder*, which helped create a genre of pop music podcast; *The Longest Shortest Time*, one of many podcasts about parenting; and the *Modern Love* podcast from the *New York Times*, a beloved show about contemporary romantic relationships.

Shows dealing with race and the American experience were ramping up by 2017 as well, an increasing number of them by and for communities of color, including *Code Switch*, *The Stoop*, *United States of Anxiety*, and *Historically Black*. In that same year, longtime producers from the public radio world began to release projects they considered experimental, in all kinds of venues—*Serial*'s *S-Town* was released that year; a number of Radiotopia's "Radiotopia Presents" shows began as well, including Al Letson's *Errthang*; Pineapple Street submitted smart and stylish work, like *Ponzi Supernova*; Marketplace's investigative podcast *The Uncertain Hour* got rolling; KCRW submitted to the archive its releases of the strangely beautiful *Bodies*, about medical mysteries, and *Everything is Alive*, about the lives of inanimate objects. A wide variety of podcast submissions in those years can be seen as responses to the Trump presidency, either explicitly (*Trump, Inc.* or *Scene on Radio*'s "Seeing White" series) or implicitly (Gimlet's *UnCivil*), and a number of shows focused on intersectionality and identity at a time of increased deportation, resurgent patriarchy, and a more overt form of White supremacy—*Afroqueer*, *Parenting While Deported*, *My Indian Life*, and *She Says*. In 2019 the *1619 Project* podcast would become the highest profile program doing historical work on race, and a few years after this sample, shows like *La Brega* and *This Place* would bring greater diversity to the podcast space from Latinx and Indigenous podcasters competing for this award.

Of course, a chronology of a few podcasts from a competition archive is an imperfect way of looking at the late 2010s. It misses a vast amount of work by independent creators, for whom contest fees or lack of awareness about these types of competition may have discouraged submission in these years, particularly for non-American and non-Anglophone podcasts not already connected to public radio producers and distributors, for whom yearly application was routine. Using an archive like this can also make it seem like winners deserve a different status to also-rans, but this is the wrong approach. The Peabodys could and did elevate investigative and experimental podcasts like *74 Seconds* and *Have You Heard George's Podcast?*, but it could also ignore perennial entrants like WNYC's *Nancy*, which would be an essential podcast to any account of changing queer experience in the late 2010s. The exercise also does not reflect very popular podcasts essential to the texture of the decade, like *The Joe Rogan Experience* or *Call Her Daddy*, which had cultural impact but little need to seek this kind of peer recognition.

These limitations aside, I have three qualitative observations from the forms that the rise of audio production took and its rapid flip from self-designating as "radio" to "podcasting." First, I think it is possible that many creators activated the term "podcast" in how they presented their work largely as a strategic choice, to capitalize on next-big-thingness and thereby acquire the resources they needed to make their work and ultimately to put it forward. The term could be used as an instrument and not merely as a category. Second, these projects were frequently narrative in character, particularly in storytelling, true crime, audio drama, and historical subject areas. Work in one or more of these categories was submitted by the most significant Peabody-submitting production companies, stations, and publishers who distributed podcasts regularly in these years and are listed in the archive. Many distributors offered works in all of these narrative forms. Categorization was also always changing—while the genre of "true crime" was reasonably stable, other bodies of material did not at first have categories, making them less discoverable. Audio drama was, for example, classified as "arts" for years by Apple before getting its own category in 2019. This leads to my final point. While trends in production are visible from a distance, the late 2010s was not a homogeneous moment. From year to year, almost from month to month, the values, subject areas, themes, funding, workflow, personnel, priorities, and techniques of modern podcasters were changing. Narrative remained a constant goal, but what did not remain constant is what "narrative" meant, how to accomplish it, or who it was for. Narrative arose at a time in which audio was undergoing a churn whose scale and pace had not been seen in a generation, and I hypothesize that it had a particular role in embodying and reflecting upon that churn. The force of the proleptic imaginary drove narrative formats into the public ear, and at the same time narratives offered an apparatus for conceptualizing the anxieties and anticipations behind this force.

Podcasting after Obsession

Now that the approach of the book and some of its conceptual parameters have been outlined, it is time to turn to the main ideas I have been alluding to in this introduction. I argue that there were three governing concepts in the narrative form of American podcasting in the late 2010s: a concept of affect that takes the form of what I call *obsession mapping*; an emergent set of experiments in narrative-based epistemology that I call *structures of knowing*; and, finally, an approach to memory that is transfixed by forgetting, what I call an *aesthetics of amnesia*. Each of the main chapters takes up one of these topics, laying out these ways of understanding podcast forms and content through a variety of examples and the discourses and events that surrounded them. While the book is intended to be primarily descriptive—at times, I have thought of this as a project whose ideal reader is a future historian searching for an attempt at conceptualizing "podcasting" after it had begun to cool down—I believe that the fact that these three features became prominent all at the same time, and in an entwined way, is no coincidence. All three connect to the ways narrative has historically been called to discursively reimagine a medium that understands itself to be in a state of identity formation. Podcasters, audiences, and critics were in these years not sure what kinds of feeling, thought, and memory were appropriate to the new medium, and their answers were both remarkably consistent (there is a startling uniformity to the structures that this book explores) and relatively weak, leaving most issues leading to them largely unresolved. This book helps to explain "how narrative worked" in the sense that its "job" seemed to be to make the medium articulate in its own unique emotional, cognitive, and mnemonic affordances. By so doing, narrative helped to reinforce the notion that the new medium indeed *had* unique affordances.

The subject of the first main chapter gives this book its title. In chapter 1, I look at a phenomenon I call obsession mapping in podcasts of the late 2010s; wherever one listens in podcasts of this unique period, especially in *Serial* and the true crime podcasts that followed it, there are narrative formulas in which podcasters conspicuously confess or mediate various forms of engrossed, almost monomaniacal, attachment to the putative objects of their programs, a tendency that had great implications for the form of podcasts and the culture that surrounded them. It was a period in which podcasting, in short, was obsessed with obsession. The culture featured an affective interplay in which obsession seemed to be the quarry that creators were after in their narratives, as it also seemed to be what listeners were after in their listening desires, a back-and-forth that is all the more apparent now that this vogue seems to have passed. In chapter 1, I outline some of the chief forms of this obsession, including the ways that the feeling is mapped to various parts of podcast structure—in the voice of its main speaker, or in the voice of another, at the outset of a podcast or as it proceeds—as well as explaining how it came to be ritualized and how podcasters employed surrogate obsessives to generate narrative tones by mimetic means. I also consider how obsession might be used as a critical strategy by scholars; one question about obsession podcasting is how it registers in popular attitudes toward the genre and how a critical approach to the work might resist responding to that registration with reflexive demystification and distance. In the conclusion of the chapter, I highlight perhaps the enduring role of obsession podcasting as it passed from vogue, in that it served as one of the first aesthetic responses to a deep problem of how a podcast narrative ought to be presented as a structured temporality.

In the second chapter, I move from how podcasts thought about feeling to how they felt about knowledge. Taking as my starting point some of the axioms by which post–*This American Life* radio storytelling and journalism told stories that focused on journalists who take us on intimate journeys to experience epiphany and empathy, I show how concerns and second thoughts about the artistic merit and even the politics of this approach grew in the late 2010s across a variety of shows. What emerged then was a shift from empathy to epistemology as one of the guiding issues of the period, as podcasts began to exhibit dramatic frustration surrounding a desire to know something that presented as difficult to ascertain. In approaching what I call the epistemological turn in podcasting studies, I use the concept of structures of knowing to explore three types of podcasts typical of this period. Some structures of knowing focused on the "for whomness" of a narrative, exhibiting a heightened sensitivity to speaking to those whose experience fell out of what had been conceptualized as normative (i.e., White, middle-class, cis audiences) public radio audiences. Many radio stories presented themselves as not fully *for* that audience and invented structures for dramatizing stark differences in "for whomness" in the place of the narrative. A second way of knowing formed around investigative podcasts and narratives of place; in these cases, the use of impasses of knowledge became the primary way in which a story set up its acts, cliffhangers, and revelations. In this way, states of knowing and not knowing could play a crucial role in the internal rhythms of long-form podcasting, something that was unavailable to the typically non-episodic anthology programs of the past. Finally, chapter 2 focuses on perhaps the most unusual subgenre of stories in this period, podcasts in which it seems as if knowledge is constantly pulling away from the listener, a pattern I call recessive epistemology. These podcasts are often investigations that take place in hinterlands and involve mysteries that not only have no offered solution but may not even be mysteries at all. Dramas of recessive epistemology have a tight set of aesthetic similarities and also the interesting property of breaking with a positivist approach to audio journalism, suggesting a world far more porous and unknowable than the one that older versions of documentary and crime programming had presumed.

In the third chapter, I turn to the cluster of podcasts that call themselves audio fiction, audio drama, scripted series, or podcast fiction, among other terms. Here I focus on what I call the "aesthetics of amnesia" that characterized this body of material. This is meant to describe the prevalence of themes of forgetting and mismemory in many prominent pieces in the period but also to name how, as a whole, the practice of making audio drama is bound up in forgetting its relationship to a denigrated past of radio-based drama. Despite participating in a medium that was nearly a century old, dramatists in the late 2010s were keen to think of themselves as breaking entirely fresh ground. The discourse produced a paradoxical sense of "radio drama" as a medium trapped as if in amber, a practice lost entirely in a distant golden age, with "audio drama" as something separate and only getting started. To critics and creators, drama was the vestige of a culture that evaporated into thin air in the 1950s—*The Shadow*, *Fibber McGee and Molly*, *Suspense*, *Dimension X*—yet at the same time it seemed also in a constant state of being invented, with its great masterpieces still ahead of us. In this way, audio drama as a field was understood both to be ready for a rebirth and also to have never been born at all, and rarely as venerably, mature, or established; it had too much past or none whatsoever. In the context of this paradox, the aesthetics of amnesia became so rich and productive, a binding concept threaded through many works of the period. "New" audio fiction would often sound new without knowing it, but it would also often sound old without knowing it. In chapter 3, I chart the ebb and flow of amnesia and its removal in podcasts ranging from *The Truth*, *Serendipity*, and *The Black Tapes* to *Limetown*, *Homecoming*, and *The Shadows*, trying to explain the ways in which these podcasts seem, on one level, to do everything they can not to look backward and yet, on another level, seem to do little else.

In approaching their individual topics, these chapters also each undertake a subsidiary project that is directly connected to their theme and is intended to engage with the question that many critics have when it comes to this medium: How should a scholar respond to a podcast, and what should a "critical" listening consist of? In my view, listening strategies prove richest when anchored in some aspect of the material itself. Because chapter 1 focuses on the theme of obsession, I ask in that context what an "obsessive" approach to reading a podcast might look like, modeling such a strategy when it comes to the first season of *Serial*. In chapter 2, where the focus is on how podcasts show vying epistemological structures, I reflect on how digital tools and other affordances can help encourage a scholarly comportment that recognizes the multiple ways in which we know what a podcast is and how it works. To know something about a podcast requires actively questioning *how* it is you know so. Finally, in chapter 3, in considering fiction podcasts and their tendency to tell stories about amnesia, mismemory, and nostalgia, while at the same time actively disavowing their own "memory" of preceding works in the same medium and genre, I show how some of these pieces can be rethought with the works of the past in mind and how works of the past can by the same token be rethought with modern audio in mind. In this way, the work of memory falls not only to the dramatist but to the critic.

As an alternative to a conclusion aimed at speculating on the future of the medium, something that this book is avoiding, I end this book with a look at one of the many things about the audio world that the overheated discourse on podcasting ignored in the 2010s: the fact that in addition to being a golden age for podcasting, this very same period has also seen a renaissance for radio art, one in which artists used the electromagnetic space of transmissions to engage some of the most complex and difficult issues facing society in the twenty-first century. I will show how radio artists explored many of the very same themes that podcasts did, but did so in ways that were often richer, more layered and challenging. One ironic result of the podcasting age is that it helped many radio artists come more directly to terms with the nature of their own craft, to embrace radio's contingency, forms of dissemination and entropy, precisely the features that tended to melt from internet-based audio. Radio art succeeded in these years, in retrospect, in part as podcasting's other, even though artists saw themselves as facing an era in which AM, low power, and even FM stations were diminishing in number. Many radio artists did with scarcity what podcasters could not do with plenty. While podcasters were preoccupied with obsession, felt called to restructure their epistemological underpinnings, and both forgot and remembered their roots, radio artists were revealing transmission ecologies hidden in the fabric of everyday life, reminding us of what made the radio band and other parts of the dynamic electromagnetic spectrum such an inherently strange place all along. To understand podcasting in the age of obsession, we must reckon with a radiophonic unconscious that it subordinated and to which it may someday return.

Audio Works Discussed in Order of Appearance in the Text

Serial (Sarah Koenig, Julie Snyder, et al., WBEZ Chicago, October 3, 2014–present, https://serialpodcast.org/).

Welcome to Night Vale (Jeffrey Cranor, Joseph Fink, Night Vale Presents, June 15, 2012–present, https://www.welcometonightvale.com/).

Last Podcast on the Left (Ben Kissel, Marcus Parks, Henry Zebrowski, Last Podcast Network, March 29, 2011–present, https://www.lastpodcastontheleft.com/).

Benjamen Walker's Theory of Everything, "Secret Histories of Podcasting" (Benjamen Walker, Radiotopia, October 21, 2015, https://theoryofeverythingpodcast.com/2015/10/secret-histories-of-podcasting/).

Open Source (Christopher Lydon, Open Source, 2003–present, https://podcasts.apple.com/us/podcast/open-source-with-christopher-lydon/id73330619?mt=2).

Radio Drama Revival (Fred Greenhalgh, January 22, 2007–present, https://radiodramarevival.com/).

Lore (Aaron Manke, Grim & Mild Entertainment, March 19, 2015–present, https://www.lorepodcast.com).

Dirty John (Christopher Goffard, *Los Angeles Times*, Wondery, October 2, 2017–November 21, 2018, https://podcasts.apple.com/us/podcast/dirty-john/id1272970334).

The Bright Sessions (Lauren Shippen, Atypical Artists, November 1, 2015–November 24, 2021, https://www.thebrightsessions.com/listen).

Lost in Larrimah (Caroline Graham, Kylie Stevenson, *The Australian*, April 27, 2018–April 15, 2022, https://podcasts.apple.com/us/podcast/lost-in-larrimah/id1377413462).

This American Life (Ira Glass, WBEZ Chicago, November 17, 1995–present, https://www.thisamericanlife.org/archive).
The Q&A with Jeff Goldsmith (Jeff Goldsmith, Unlikely Films, Inc., March 16, 2011–present, http://www.theqandapodcast.com/).
The Bill Simmons Podcast (Bill Simmons, The Ringer, October 1, 2015–present, https://podcasts.apple.com/ca/podcast/the-bill-simmons-podcast/id1043699613).
WTF with Marc Maron (Marc Maron, Boomer Lives! Productions, September 1, 2009–present, http://www.wtfpod.com/podcast).
Radiolab (Jad Abumrad, Robert Krulwich, WNYC, May 25, 2012–present, https://radiolab.org/episodes).
L.A. Theatre Works (Susan Albert Lowenberg, LATW, February 9, 2017–present, https://latw.org/podcasts).
Reveal (Al Letson, Center for Investigative Reporting, Public Radio Exchange, September 18, 2013–present, https://revealnews.org/episodes/).
Snap Judgment (Glynn Washington, Snap Judgment Studios, PRX, July 2010–present, https://snapjudgment.org/podcast-episodes/).
Radiolab Presents: G (Pat Walters, WNYC Studios, June 7, 2019–July 15, 2021, https://radiolab.org/series/radiolab-presents-g).
More Perfect (Julia Longoria, WNYC Studios, May 24, 2016–October 8, 2020, https://www.wnycstudios.org/podcasts/radiolabmoreperfect).
Dolly Parton's America (Jay Abumrad, Shima Oliaee, OSM Audio, WNYC Studios, October 3, 2019–December 31, 2019, https://www.wnycstudios.org/podcasts/dolly-partons-america).
Storycorps (David Isay, WBEZ Chicago, 2003–present, https://storycorps.org/podcast/).
Nocturne (Vanessa Lowe, October 20, 2014–present, https://nocturnepodcast.org/nocturne-episodes/).
The Moth Radio Hour (Jay Allison, Meg Bowles, George Dawes Green, Jenifer Hixson, Suzanne Rust, The Moth, Atlantic Public Media, Public Radio Exchange, 2009–present, https://themoth.org/radio-hour).
The Lapse (Kyle Gest, The Lapse Storytellers, February 8, 2014–October 16, 2017, https://podcasts.apple.com/us/podcast/the-lapse-storytelling-podcast/id816816401).
Everything Is Stories (Mike Martinez, Tyler Wray, August 26, 2013–December 11, 2018, https://podcasts.apple.com/us/podcast/everything-is-stories/id788131884).
Comedy Bang! Bang! (Scott Aukerman, Earwolf, Midroll Media, May 1, 2009–present, https://www.earwolf.com/show/comedy-bang-bang/).
Freakonomics Radio (Stephen J. Dubner, WNYC Studios, February 5, 2010–present, https://freakonomics.com/series/freakonomics-radio/).
Pop Culture Happy Hour (Aisha Harris, Linda Holmes, Stephen Thompson, Glen Weldon, NPR, July 22, 2010–present, https://www.npr.org/podcasts/510282/pop-culture-happy-hour).
Hidden Brain (Shankar Vedantam, Hidden Brain Media, NPR, September 21, 2015–present, https://hiddenbrain.org/category/podcast/).

Invisibilia (Lulu Miller, Alix Spiegel, NPR, January 9, 2015–present, https://www.npr.org/programs/invisibilia/).

The New Yorker Radio Hour (David Remnick, WNYC Studios, October 23, 2015–present, https://www.wnycstudios.org/podcasts/tnyradiohour).

2 Dope Queens (Phoebe Robinson, Jessica Williams, WNYC Studios, April 5, 2016–March 12, 2019, https://www.wnycstudios.org/podcasts/dopequeens/episodes).

Death, Sex & Money (Anna Sale, WNYC Studios, May 4, 2014–present, https://www.wnycstudios.org/podcasts/deathsexmoney/episodes).

99% Invisible (Roman Mars, KALW Public Media, Radiotopia, September 23, 2010–present, https://99percentinvisible.org/archives/).

Ear Hustle (Nigel Poor, Earlonne Woods, Rahsaan "New York" Thomas, Radiotopia, June 14, 2017–present, https://www.earhustlesq.com/listen).

The Heart (Kaitlin Prest, Mermaid Palace, Radiotopia, February 5, 2013–February 23, 2022, https://www.theheartradio.org/all-episodes).

The Memory Palace (Nate DiMeo, Radiotopia, November 12, 2008–present, https://thememorypalace.us/episodes/).

Reply All (Alex Goldman, PJ Vogt, Gimlet Media, November 24, 2014–June 23, 2023, https://gimletmedia.com/shows/reply-all).

Heavyweight (Jonathan Goldstein, Gimlet Media, September 23, 2016–present, https://gimletmedia.com/shows/heavyweight#show-tab-picker).

Pod Save America (Jon Favreau, Jon Lovett, Dan Pfeiffer, Tommy Vietor, Crooked Media, January 9, 2017–present, https://crooked.com/podcast-series/pod-save-america/).

Revisionist History (Malcolm Gladwell, Pushkin Industries, June 16, 2016–present, https://www.pushkin.fm/podcasts/revisionist-history).

Breakdown (Bill Rankin, *Atlanta Journal-Constitution*, 2014–present, https://www.ajc.com/news/breakdown/).

Unsolved (Gina Barton, *Milwaukee Journal-Sentinel*, 2015–present, https://projects.jsonline.com/topics/unsolved/season-three/the-devil-you-know.html).

Criminal (Phoebe Judge, Radiotopia, 2014–present, https://thisiscriminal.com/).

Crimetown (Marc Smerling, Zac Stuart-Pontier, Gimlet Media, 2016–19, https://gimletmedia.com/shows/crimetown/episodes#show-tab-picker).

16 Shots (Jenn White, WBEZ Chicago, 2018–22, https://www.wbez.org/shows/16-shots/).

The Daily (Michael Barbaro, Sabrina Tavernise, New York Times Company, 2017–present, https://www.nytimes.com/column/the-daily).

Bundyville (Leah Sottile, Longreads, Oregon Public Broadcasting, April 25, 2018–July 15, 2019, https://www.opb.org/news/article/bundyville-occupation-podcast/).

The Catch and Kill Podcast with Ronan Farrow (Ronan Farrow, Pineapple Street Studios, November 26, 2019–July 26, 2021, https://podcasts.apple.com/us/podcast/the-catch-and-kill-podcast-with-ronan-farrow/id1487730212).

The FRONTLINE Dispatch, "The Weight of Dust" (Sophie McKibben, Michelle Mizner, PBS, December 13, 2018, https://podcasts.apple.com/us/podcast/the-weight-of-dust/id1277250997?i=1000425709770).

The Truth (Cadence Mandybura, Jonathan Mitchell, Radiotopia, February 11, 2012–present, http://www.thetruthpodcast.com/).

The Thrilling Adventure Hour (Ben Acker, Ben Blacker, Forever Dog Podcast Network, January 23, 2011–present, https://podcasts.apple.com/us/podcast/the-thrilling-adventure-hour/id408691897).
PleasureTown (Keith Ecker, Erin Kahoa, WBEZ, June 13, 2014–December 7, 2016, https://www.wbez.org/shows/pleasuretown/)
Limetown (Zack Akers, Skip Bronkie, Two-Up Productions, July 29, 2015–December 3, 2018, https://twoupproductions.com/limetown/podcast).
The Black Tapes (Paul Bae, Terry Miles, Pacific Northwest Stories, May 21, 2015–January 12, 2020, http://theblacktapespodcast.com/).
The Message (GE Podcast Theater, Panoply, October 3, 2015–November 21, 2015, https://podcasts.apple.com/us/podcast/lifeafter-the-message/id1045990056).
Our Fair City (Clayton Faits, Jeffrey Gardner, HartLife NFP, March 28, 2012–July 4, 2018, https://podcasts.apple.com/us/podcast/our-fair-city/id514748675).
Wolf 359 (Zach Valenti, Gabriel Urbina, Sarah Shachat, Kinda Evil Genius Productions, August 15, 2014–December 24, 2017, https://wolf359.fm/season1).
The Bright Sessions (Lauren Shippen, Atypical Artists, November 1, 2015–November 24, 2021, https://www.thebrightsessions.com/listen).
Alice Isn't Dead (Joseph Fink, Night Vale Production, March 7, 2016–November 14, 2019, http://www.nightvalepresents.com/aliceisntdead).
Within the Wires (Jeffrey Cranor, Janina Matthewson, Night Vale Production, June 20, 2016–present, http://www.nightvalepresents.com/withinthewires).
The Unexplainable Disappearance of Mars Patel (Benjamin Strouse, Pinna and Gen-Z Media, June 20, 2021–July 23, 2021, https://gzmshows.com/shows/listing/mars-patel/).
Homecoming (Eli Horowitz, Gimlet Media, November 16, 2016–November 2, 2018, https://gimletmedia.com/shows/homecoming/episodes).
36 Questions (Chris Littler, Ellen Winter, Two-Up Productions, July 10, 2014–August 7, 2017, https://twoupproductions.com/36-questions/podcast).
Issa Rae Presents . . . Fruit (Issa Rae, Issa Rae Productions, Stitcher, February 2, 2016–July 17, 2017, https://podcasts.apple.com/us/podcast/issa-rae-presents-fruit/id1256093382).
Damned Spot (Eric Nuzum, Audible Originals, October 26, 2017, https://www.audible.com/pd/Damned-Spot-Podcast/B08DDF1FY8).
Where Should We Begin? With Esther Perel (Esther Perel, Audible Originals, October 24, 2017, https://www.audible.com/pd/Where-Should-We-Begin-with-Esther-Perel-Podcast/B08DDCJ44L#:~:text=Step%20into%20iconic%20relationship%20therapist,to%20be%20heard%20and%20understood).
Song Exploder (Hrishikesh Hirway, Radiotopia, January 1, 2014–present, https://songexploder.net/episodes).
The Longest Shortest Time (Hillary Frank, Stitcher, December 24, 2020–December 17, 2019, https://longestshortesttime.com/episodes).
Modern Love (Anna Martin, New York Times Company, December 7, 2015–present, https://www.nytimes.com/column/modern-love-podcast).
Code Switch (Gene Demby, Shereen Marisol Meraji, NPR, May 9, 2016–present, https://www.npr.org/podcasts/510312/codeswitch).
The Stoop (Leila Day, Hana Baba, Radiotopia, July 18, 2017–present, http://www.thestoop.org/).

The United States of Anxiety (Kai Wright, WNYC Studios, September 18, 2018–August 13, 202, https://www.wnycstudios.org/podcasts/anxiety/archives).
Historically Black (Tracy Clayton, Roxane Gay, Keegan-Michael Key, Heben Nigatu, Issa Rae, APM Reports, *Washington Post*, September 19, 2016–November 7, 2016, https://features.apmreports.org/historically-black/).
S-Town (Brian Reed, Serial Productions, March 28, 2017, https://stownpodcast.org/).
Errthang (Al Leston, Radiotopia, May 2015–January 26, 2019, https://radiotopiapresents.fm/errthang).
Ponzi Supernova (Steve Fishman, Audible, May 5, 2017–June 9, 2017, https://ponzi.libsyn.com/).
The Uncertain Hour (Krissy Clark, Marketplace, April 28, 2016–March 24, 2021, https://www.marketplace.org/shows/the-uncertain-hour/).
Bodies (Allison Behringer, KCRW, July 25, 2018–December 1, 2021, https://www.kcrw.com/culture/shows/bodies).
Everything Is Alive (Ian Chillag, Radiotopia, July 16, 2018–present, https://www.everythingisalive.com/).
Trump, Inc. (Andrea Bernstein, Ilya Marritz, ProPublica, WNYC Studios, October 5, 2018–January 19, 2021, https://www.wnycstudios.org/podcasts/trumpinc).
Scene on Radio, Season 2: "Seeing White" (John Biewen, Center for Documentary Studies at Duke University, February 15, 2017–August 24, 2017, https://www.sceneonradio.org/seeing-white/).
UnCivil (Jack Hitt, Chenjerai Kumanyika, Gimlet Media, September 20, 2017–October 27, 2020, https://gimletmedia.com/shows/uncivil).
Afroqueer (Selly Thiam, AQ Studios, July 19, 2018–June 17, 2022, https://afroqueerpodcast.com/episodes/).
The New Yorker Radio Hour, "Parenting While Deported" (David Remnick, *New Yorker*, WNYC Studios, September 7, 2018, https://www.wnycstudios.org/podcasts/tnyradiohour/episodes/parenting-while-deported).
Kalki Presents: My Indian Life (Kalki Koechlin, BBC World Service, August 3, 2018–present, https://www.bbc.co.uk/programmes/p06dwm6y/episodes/downloads).
She Says (Sarah Delia, WFAE, May 16, 2018–December 4, 2019, https://www.npr.org/podcasts/611786868/she-says).
1619 (Nikole Hannah-Jones, New York Times Company, August 23, 2019–October 11, 2019, https://www.nytimes.com/2020/01/23/podcasts/1619-podcast.html).
La Brega (Alana Casanova-Burgess, WNYC Studios, Futuro Studios, February 24, 2021–present, https://www.wnycstudios.org/podcasts/la-brega).
This Place (Rosanna Deerchild, CBC Radio, June 17, 2021–August 24, 2021, https://www.cbc.ca/listen/cbc-podcasts/1020-this-place).
74 Seconds (Jon Collins, Riham Feshir, Tracy Mumford, MPR News, May 22, 2017–August 14, 2017, https://www.mprnews.org/podcasts/74-seconds).
Have You Heard George's Podcast? (George the Poet, BBC Radio, August 31, 2019–September 15, 2021, https://www.bbc.co.uk/programmes/p07915kd/episodes/downloads).

Nancy (Tobin Low, Kathy Tu, WNYC Studios, April 9, 2017–June 29, 2020, https://www.wnycstudios.org/podcasts/nancy/episodes).
The Joe Rogan Experience (Joe Rogan, Spotify, December 23, 2009–present, https://open.spotify.com/show/4rOoJ6Egrf8K2IrywzwOMk).
Call Her Daddy (Alexandra Cooper, Barstool Sports, October 3, 2018–present, https://open.spotify.com/show/7bnjJ7Va1nM07Um4Od55dW

CHAPTER I

Obsession, from Map to Strategy

Obsessed with *Serial*

In November 2014, New York comedy duo Sal Gentile and John Purcell, who performed together as Cartwright, posted a video on YouTube about a true crime show then single-handedly transforming how podcasts were characterized in the United States and around the world.[1] In the comic skit, Gentile and Purcell sit in a dim living room, drinking from oversized mugs, the liquid never reaching their lips, their eyes cast aside in contemplation, lampooning the style you might see in novice filmmaking. Gentile gives a toss of the head: "John, do you know that new podcast *Serial*? I am obsessed with it." John, a little aggressive, doesn't hesitate. "Oh my God, I love that podcast so much," he says. "That is all I can talk about with any of my friends right now." John mentions a subreddit forum in which fans were posting theories about the show. Sal has read every post in the subreddit, he insists, pointing out there was also a subreddit about the subreddit discussing what was really going on in the subreddit. Not to be outdone, John says he follows a second podcast by *Slate* on discussions about *Serial*. Sal swears he has heard every episode of that *Slate* podcast and even follows a podcast that's about the *Slate* podcast about *Serial*. John notes his fondness for an episode of the *Slate* podcast about *Serial* that reenacted a reenactment of a murder timeline on an episode of *Serial*; Sal recalls the time when the podcast about the podcast about the podcast re-created the re-creation of the re-creation.

There is a pause. "Oh my God, I'm obsessed with this conversation. I mean the one we're having," says John, pointing offscreen, "and you should totally check out this conversation just over here where they're presenting theories about what they think we're talking about." Sal is way ahead of him: "I'm obsessed with that conversation." The camera pans to two others at a table in earshot, also pretending to drink tea, discussing how much they love the conversation that Sal and John are having—and the subreddit about it. The day that this video was released on the Funny or Die website, *Serial*'s Facebook page reposted it. "Oh my God," the post read, completing the mise en abyme, "we're obsessed with this video about how people obsess over *Serial*."[2]

It is surely not the first good joke about a podcast, but it may be the first good joke about a cultural style that surrounded podcasting during the years between *Serial* and the COVID-19 pandemic, a style at the heart of this chapter. Like the Cartwright joke (and the reactions of the people in it), the experience of podcast aesthetics would over the course of the next few years often get caught up in the work of admiring the scale of its seemingly bottomless recursiveness. "I think of radio as a buddy," Laura Starecheski said in a 2014 interview of prominent radio producers for the Transom website, "and a podcast is like a wormhole that you get sucked into."[3] *Serial* as a phenomenon involved Facebook posts about YouTube videos about conversations about subreddits about fan podcasts, with the program itself at the center of many hall-of-mirror discourses whose horizon nobody could see. Besides the number of downloads of the program—some 90 million within the first year (powered by the fact that Apple preinstalled its podcast app on its operating system at that same time) and nearly 420 million by the time *Serial* was folded into the *New York Times* podcasting outfit four years after that—it was the show's sheer social media omnipresence that got advertisers and publishers interested in podcasting, sparking talk of a kind of golden age dawning.[4] Whether or not downloads translated into actual listening was unclear, even to executives in the industry at the time, and it would still be three years before Apple released analytics to podcast publishers, allowing them to glimpse how far into an episode listeners actually listened.[5] Indeed, despite *Serial*'s record of downloads, Edison Research found in its 2015 annual survey that just 10 percent of podcast listeners had even heard of the show.[6] *Serial*'s social media footprint, by contrast, was easy to measure—Erica Haugtvedt found some fifty thousand participants in the *Serial* subreddit in March 2016—leaving behind an unmatched corpus of public ensorcellment.[7]

This is why, while podcasts themselves are not "social" media, as they are experienced individually and are onerous to recirculate in morsel form across networks of users, many writers have found it hard to speak of podcasting without recognizing how the rise of social media in the same years as the golden age of podcasting provided a framework in which the latter medium could grow, nourished by symbiotic tweets, feeds, listicles, recommendations, newsletters, rumors, clubs, slacks, and videos.[8] Many key texts in podcast studies trace this activity meticulously, from Lance Dann and Martin Spinelli's studies of podcaster social media use to Sarah Florini's analysis of podcasts as part of a trans-platform network able to empower African Americans to navigate racialized life in America.[9] There is an important commercial side to this circulation, as well. Cartwright's skit highlights that the podcast form often generates urgency for redistribution across interpersonal relationships ("It's all I can talk about with my friends!"), something that podcasters would learn to channel into much-needed revenue by creating Patreons and other direct funding streams, subscriptions, discussion forums, stickers, T-shirts, and coupon codes for sponsors that enlisted fans to the task of keeping content circulating in an online loop, in this case one that leads all the way back to *Serial*'s own newsfeed. Like embeddedness in social media, the crowdfunding dimension of podcasting is among the features that led observers to argue that podcasting constituted a "new medium," although a case could be made that these features had obvious roots in linear radio, on a different scale. The limited-edition coin sent to exclusive circle Patreon supporters by an upstart podcast surely has an ancestor in the public radio station's annual fundraising drive tote bag.

To launch this chapter, I would like to focus on how the Cartwright joke also has its finger on something else, a mode of comportment that many late 2010s podcast works shared and one that I will explore in this chapter at the textual level: obsession. When *Serial* became a hit, it was celebrated everywhere—*Buzzfeed*, *Forbes*, *Rolling Stone*, even the *Colbert Report* and *Saturday Night Live*—not just for what it was but for how it felt, making listeners insatiably connected to its story.[10] For many writers, that insatiability called for a discursive articulation, and into the open door of this requirement stepped the argot of obsession, which helped fans and critics characterize their affective relation to the piece. In the *New York Times*, critic Ernesto Londoño insisted it was the subject matter of the podcast that leant itself to "obsession," while WHYY interviewer Terry Gross felt that listeners were more "obsessed" with what the host, Sarah Koenig, would conclude at the end of the show rather than with these events themselves.[11] *The Guardian*'s Miranda Sawyer confessed her own "obsession" with the show, taking on the language of pop therapy: "I need you all to start listening so that we can form some kind of *Serial* support group."[12] *Vulture*'s Dan Fitchette insisted you couldn't "swing a beard in Brooklyn" without hitting a *Serial* fan and wrote a listicle to spread the word ("5 Reasons Everyone's Obsessed with 'Serial'"), while a year later *Vanity Fair*'s James Wolcott took an opportunity for some self-mockery ("So, Like, Why Are We So Obsessed with Podcasts Right Now?").[13] In a time when the argot of addiction was often used to frame TV (the terms "binge watching" and "addictive TV" grew popular in these years), podcasting was associated with something else: deep fixation, persistent passion. Over time, in the way of infatuation, the feeling flattened. A 2021 piece by *Vice* reviewed burgeoning psychological literature about how listening to true crime podcasts could lead to sleep deprivation that resulted in high blood pressure, diabetes, obesity, and depression. "If your true crime habit is inhibiting your ability to leave the house, get work done, or live without an overwhelming sense of paranoia," the article read, in a solicitous tone uncharacteristic of the irreverent news brand, "experts suggest seeking help from a mental health professional."[14]

For the purposes of this chapter, I propose anchoring the theory of "obsession" that subtends this discourse in two ways. The first is historically proximal: obsession is to some extent the evolution of an idea of "interestingness" that public radio producers, especially those associated with *This American Life*, often prized in the years prior to *Serial*. The task of "finding interestingness" is a deep impulse in American public radio's idea of "good writing," a talisman pursued by training, critique, and editing. Consider Nancy Updike's seminal 2006 Transom essay "Better Writing Through Radio," in which she uses the term eleven times, often in language linked to the adjectives of vivid body experience: expect too much of the tape and you get "flabby, barely-interesting radio"; a radio story has to be "sluttier with its charms" rather than being coy about the "most interesting" stuff; to write "interestingly" you need to control the "draggy parts" of a narrative.[15] To acquire peak radio fitness, so to speak, is to find and to adhere to a strict diet of interestingness.

Jessica Abel's book *Out on the Wire* also latches onto the word, with its many interviews with *TAL* figures, including host Ira Glass and producers such as Sean Cole and Julie Snyder. In that text—created in the form of a graphic novel—Glass is shown speaking of making radio as a task of "amusing oneself" after having cultivated a self-aware sort of taste. "Interesting" appears a half dozen times in this advice-driven passage, arguing that the "hook" of any story is said to be whatever is "interesting" about its topic, something that a producer has to find or create. The interesting is best when it seems to come naturally, reflecting a preternatural instinct and keen intuition. A classic example Abel provides is producer Alex Blumberg's story about the roots of the 2008 financial crisis, "The Giant Pool of Money," which, according to the book, developed out of Blumberg's "obsession" with the housing market around the time of the financial collapse, before developing into an admirable audio piece in which a surfeit of global wealth was itself the "main character." Snyder and Updike (as producers and editors of *Serial*) and Blumberg (as cofounder of Gimlet Media) were in an ideal position to bring this sensibility to the center of modern podcasting taste.[16]

Of course, "interestingness" as an aesthetic judgment is mystified by its entanglement with a romantic notion of an intuition without origin, the secular version of divine inspiration. "The interesting" carries an immediate and deep significance to the person who feels it, yet there is a niggling opacity to that significance. It seems to differentiate one sensibility from another more than it explains some object in the world, unlike other aesthetic judgments like "the beautiful" or "the uncanny," which we typically anchor in identifiable attributes of the art object that can be disputed yet remain, on some material level, shared. Theorist Sianne Ngai has explored the question of the interesting in some depth. In her view, we judge objects to be "interesting" without immediate external criteria to point to. The "interesting," she writes, "ascribes value to that which seems to differ, in a yet-to-be-conceptualized-way, from a general expectation or norm whose exact concept may itself be missing at the moment of judgment."[17] When we find an object interesting, we know it stands out but not why, and there is something vexing about this omission, as well as the nontransparency of our own intuition. We can't help but wonder about the criteria, hidden even to ourselves, that lead us to find some whatsit we have encountered so compelling. Ngai puts it in the form of a question we ask ourselves: "What was it I must have noticed and simultaneously not noticed about the appearance of the object in order to have judged it interesting?"[18] In the context of narrative radio production, "the interesting" has a curious function, often serving as a prompt for the work of making audio. The surprise appearance of interest in a producer's affective response to some story, place, or person compels them to produce a story that can potentially explain this sharply felt response. In this way, "interestingness" is both an attribute of a thing that serves as a motive for making a radio story and a motive that becomes an attribute of the radio story about the thing.

If obsession marks a moment in the history of podcasting, it is also true that podcasting marks a moment in the history of obsession. For this reason, the second root of "obsession" I'm drawing on here is as a very old pop psychological term, one historically linked to the notion of an idée fixe, by which I mean thoughts that seem to consume us and are more than a taste, passion, or hobby. These are thoughts that interfere with patterns of life for those of us who are otherwise unaffected by mental illness. Today it is common to associate such obsession with compulsion disorders and to treat them with a combination of cognitive behavioral therapy and medication. For much of its history, the diagnosis was linked directly to trauma, particularly in the work of psychoanalyst Pierre Janet, who wrote a major two-volume study in the early twentieth century classifying some sixty varieties of idées fixes, theorizing the phenomenon primarily as a response to the draining of psychic energy that accompanies trauma and suggesting various treatments depending on how tyrannizing thoughts manifest themselves.[19] Actually, in modern intellectual history the idée fixe is even older than that account, stretching back in French psychiatry to the era of the Bourbon Restoration and the concept of "monomania." In her classic book on the nineteenth-century rise of psychiatry as a profession, historian Jan Goldstein explores monomania as the theory of "a single pathological preoccupation in an otherwise sound mind," something that set it apart from cases of patients with broad generalized manias on all topics, as a kind of halfway point between full mania and deep melancholy.[20] First used by Jean-Étienne Esquirol in 1810, the term became common in the 1820s and entered the French lexicon during the subsequent decade. According to Goldstein, monomania was a socially and politically loaded vogue, accounting for a significant fraction of diagnoses for decades and playing a role in the expansion of the "insanity" defense in courts, thanks to the efforts of evangelists like Étienne-Jean Georget, before vanishing almost entirely as a diagnosis by the twentieth century.

Monomania had a profound influence on the arts and literature, particularly in France, where it was reflected in a rise in stories of compulsive collecting, the need for enclosure, and the frantic gathering of knowledge, all hallmarks of a tradition stretching from Flaubert to the contemporary artist Sophie Calle. Critic Marina van Zuylen's reflections on this topic consider monomania to be a sort of counterattack against modern instabilities, a desire for control over the dangers of the world. By "offering a life altogether wed to an idea" we escape our anxieties over reality; in this way the idée fixe empowers as it disables at the same time.[21] That such an impulse should return with the rise of social and streaming media comes as no surprise. Perhaps the rise of monomania as a revived cultural form is little more than an obvious response to living in a media ecology in which algorithms, a kind of prosthetic monomania, frequently serve us new versions of what we already like, know, or feel. Nothing is more monomaniacal, after all, than a streaming service's recommendation list. On the one hand, monomaniacs today resemble the media they use, while on the other hand they represent an effort to wrest a modicum of uniqueness and self-possession seemingly lost to those media. When it comes to the question of "why obsession?," then, my hypothesis is that podcasts of the late 2010s sought conceptual turf by wedding a public radio idea of "the interesting" to an atavistic literary notion of the fixed idea, which had new salience as an affective reaction to some of the instability associated with the intensification of mediated life, something of which the rise of podcasting was itself one of many symptoms. If podcasting emerged partly as the product of radio responding to the logic of the internet's infrastructure, narrative podcasting was the response of public radio aesthetics to the logic of its algorithms. The rise of obsession podcasting may also stand in relation to the rise of conspiracy culture, which is a little like obsession's political mirror image in the culture of the late 2010s. The age of obsession in podcast media for the American Left was the age of conspiracy in social media for the American Right; it is not outside of the realm of possibility that an itch that *Serial* scratched for one population is the same itch that QAnon scratched for another.

Obsession culture was certainly understood as a kind of balm by critics. The most successful episode of Gimlet's flagship program, *Reply All*'s 2020 "The Case of the Missing Hit," begins with host PJ Vogt's discussion of his own diagnosis of an obsessive disorder and relates this to filmmaker Tyler Gillett's quest to tack down an earworm tune that is plaguing him, a quixotic story that critic Nicholas Quah called "legendary" and *The Guardian*'s Hannah Davies called "the best podcast episode ever."[22] "In a society that barrages us endlessly with methods for optimization and productivity," *Esquire*'s Lauren Kranc wrote, "it is both hilarious and immensely gratifying to invest in a relatively pointless, albeit meticulous and comprehensive investigation."[23] With this cultural backdrop and rough genealogy in mind, I want to call attention to how frequently obsession is at work *within* narrative podcasts like "Missing Hit," in ways that are crucial to their self-justification, mood, and poetic structure.

After all, it is not merely the fact that audience members became obsessed with the show that matters. Just as you can follow an obsession with a podcast from one medium to another, you can also follow it within a program. It is significant, for instance, that Gillett's obsession is introduced through the lens of Vogt's obsessiveness, generating a kind of map from one storyteller to another. Indeed, in what follows I show that "maps of obsession" structure many of these works themselves, offering a systematic infrastructure to how these pieces think about feeling and framing the criteria under which they demonstrate a sense of their own sophistication, depth, and merit. I would go so far as to say obsession mapping is a formula that had "controlling importance" in the podcast genre, to borrow a term from Michael Fried, in that it linked many podcasts that don't otherwise necessarily go together.[24] By mapping obsession, podcasters enriched, frustrated, and even undermined key pieces of the late 2010s period, to a point of saturation that inevitably led to a search for alternative themes. In one sense, the "era of obsession" in narrative podcasting is hardly an "era" at all: it rose sharply and then suddenly splintered and declined. Future historians might not even see it as a form with a meaningful diachronic temporal contour. Yet it is a clear signature style of the period, and it is the task of this chapter both to account for its prevalence and to gamble on its potential as a critical instinct. "Obsession podcasting" is not just a historical puzzle to sort out, after all, but also a way of describing and doing critical podcast listening.

Maybe Too Strong a Word

With the language of obsession marking public discourse about *Serial*, it should come as no surprise that this theme is also embedded in the first episode, "The Alibi," and in a curiously self-negating manner. Although the season focused on a specific story—the grisly murder of teenager Hae Min Lee, according to authorities at the hands of her ex-boyfriend Adnan Syed, after school one day in 1999—host Sarah Koenig famously backs into that story by way of reflecting on how hard it is for most people to remember what they did in recent days or weeks, let alone years ago. As a "lark" she interviews her teenage nephew Sam and his friend Tyler, who confirm her hypothesis, misremembering the events of a recent weekend. The mnemonic experiment sets into motion the question of reliable recall at the center of the narrative, but it is also a way of examining the producer's own memory, posing the question of where the story started for her. Koenig starts again, explaining how the story of the fifteen-year-old crime came to her from attorney Rabia Chaudry along with her brother Saad, whom she visits on tape for a talk about the case and its many unresolved questions. She explains: "This conversation with Rabia and Saad, this is what launched me on this yearlong—'obsession' is maybe too strong a word—let's say fascination with this case."

One of Koenig's attributes as a narrator lies in her infrequent asides that lend color to her narrative voice, serving the phatic role of establishing communication between her and her addressee. Sometimes these are humanizing windows that seem to both break and reinforce the perimeter of her studied demeanor of amiable, worldly reserve. In the seventh episode of the first season, for instance, she mentions that her father was fond of saying that "all facts are friendly"; later in that season she intimates that she lost both her father and her stepfather within months of each other. At the start of the second season, she speaks of an illustrated book she read to her kids that showed scales of events constantly expanding, which struck her as a master trope for that season's exploration of the war in Afghanistan. Other asides can be played for mirth, as in the moment in the fifth episode in which we hear tape of Koenig and producer Dana Chivvis walking through the crime's timeline, when Chivvis is distracted from Koenig's off-the-cuff theorization by the sign for a special on shrimp at a nearby restaurant. The tape is interrupted by Koenig's narration, located later but speaking in the present tense, with the jab "Sometimes I think Dana isn't listening to me." Pauses can work in a similar way. In the ninth episode of the second season, for example, Koenig mentions that in a national security context, doing a prisoner swap with the enemy is called a "mutual release." A pause and then the sound of Koenig clearing her throat, as if about to say something but then thinking better of it, demurely marks a place for the unsaid punchline of an obvious prurient joke.[25]

But the "obsession" aside is different. Consider it relative to the rest of the sentence. At the outset, Koenig provides a double use of "this" (referring to the conversation with Rabia) with vocal emphasis both times—"*this* conversation . . . *this* is what launched me." She seems *so sure* that it was this very moment a year ago that inaugurated the timeline of her own narration. The prosody conveys that this certainty has been arrived at upon reflection, a concluded thought, as if to say, "Dear listener, I have given this much consideration and put my finger on where, for me, this project started." It's a defining gesture for many internal character narrators searching for a place to begin. But then, immediately, the sentence wavers from stability to instability as the edifice of certainty crumbles. The definitive use of the verb "to be" is all of a sudden qualified by "maybe," then the even more suspicious "let's say," which makes it sound like a false story in which we are being enlisted as coconspirators. Not only does Koenig use an aside to second-guess herself, suggesting more than one ongoing stream of thought, but she uses that framework to search for a word that isn't yet known—do I mean "obsession," or do I mean "fascination?"—and although she settles on the latter, the more suspicious former is already set free to run amok in the world of the narrative, like a genie from a bottle. If obsession were truly discarded, then why would editors with the deep experience of this team write and preserve the aside in the script? And the question of obsession remains a lurking hint in the language of the show, returning now and then across its episodes. In the sixth episode, Koenig goes so far as to revise her view that it was the visit to Rabia and Saad that got her so involved in the case, admitting that "really what hooked [her] most" was Syed himself.

The opening moment also may serve a larger poetic purpose. In it, the narration re-creates in miniature what the whole season would eventually do. What drives *Serial*'s seriality, after all, is neither the chronology of the story nor that of its reconstruction but Koenig's thought process, her certainty-then-uncertainty about Syed.[26] What we are listening to is Koenig and her team organize and reorganize an ever-growing wall of stubbornly ambiguous details, and value and devalue them, recursively, incredulously, passionately. A good example comes in the eleventh episode, as Koenig wonders aloud about a story that follows Syed preventing a possible fight by kissing a friend's cheek. Does this gesture signal he is a peacemaker or that he is ice-cold? It's a classic question about what literary theorist Vladimir Propp called a "function," an act that can't be defined unless we put it in the context of a narrative that will see it in terms of other acts either earlier or later in the narrative. If we know Syed will commit the act of murder, then we know which function the kiss should have; if we know he does not commit that act, then we go the other way.[27] In the third episode, Koenig likens items at the crime scene to objects in a game of Clue; in the sixth episode, she likens her thought process to swiveling a Rubik's Cube until it begins to fit, only to see it crumble; in the seventh episode, one of her interviewees likens the process of thinking through these crimes to juggling: many of these analogies take the form of haptic games of skill and intellect that foil the novice and reveal the savant. Most importantly, each episode ends the same way, with the opening of a possibility for a reversal of certainty in which Koenig might change her mind about what really happened in 1999, the gameplay about to turn.

In previous work, I have argued that what many fans conventionally call radio's "theater of the mind" is often a theater *about* the mind, that is, a medium that deploys various models of mental processes that themselves evolve over time, making radio a place in which to perform an expressive exploration of what we mean by "mind" in the first place.[28] This podcast is an instance of exactly that. The story is set in Baltimore County, but *Serial* actually takes place in Koenig's seesawing mind, and its main action is the depiction of her open-ended back-and-forth cognizing. In this way, the season is really unlike many other investigative podcasts of its era. In the second season of *In The Dark*, the producers sift through eight courthouses to find twenty-five years of trials and 225 jury records to study a district attorney's record; in 2018 the podcast *Reveal* analyzed thirty-one million government mortgage records to prove that in sixty-one metro areas Black Americans were more likely to be denied mortgages than Whites. Both of these projects required incredible effort, but neither dramatizes the interior mental churn of the researchers as they struggle with their archives. These podcasts *feature* cognitive labor, but they are not *about* cognitive labor in the way *Serial* is. The latter is more akin to contemporaneous experimental radio features of its era, many of which explore the "dramaturgy of thoughts" in this period: Pejk Malinovski's 2013 piece "Slow Movement: Everything, Nothing, Harvey Keitel," which follows the thoughts of a man sitting next to the famous actor in a meditation; Cathy FitzGerald's 2012 "Dreaming Dickens," a docu-fantasy about slipping into Charles Dickens's consciousness while moving through London; Tim Hinman's 2014 *In One Ear and Out the Other* about a miniature explorer moving through his own ear toward the brain. Readers familiar with these works will find the comparison farfetched, but *Serial* does share the sense of a writer trying to perform vividly the tyranny of their own thought process, something that traditional radio often pushes behind the wall of the editing process. To put it another way, what *Serial* dramatizes—across all three of its seasons—is essentially a long thought process, one that takes the form of a valiant attempt to defend the naive paternal apothegm that facts are indeed friendly, achieving only mixed success.

Returning to "the Alibi," Koenig's confession of possible obsession has even greater weight because it is just the first of two suspicious denials of obsession bookending the episode. The second instance occurs as we begin to learn about Koenig's conversations with Syed in prison. Syed denies involvement in the crime, and Koenig relates his claims about how he felt about Lee. Again, in Koenig's narration, "obsession" enters in the form of a negation:

> He says he loved her in the way of high school love, but then also in the way of high school love got over her. So that when they broke up for good sometime before Christmas break of senior year, he says he was sad for sure, but not obsessed or anything.

I would like to highlight two dimensions of this brief but important moment. First, although it links Koenig and Syed, the term "obsession" is used differently here, referring not to a journalist's overeagerness that might cloud her reason but instead to the theory (put forward at trial) that Syed had a pathology that led him to be possessive and controlling of Lee, as their relationship went against his religious values. It is the sort of obsessiveness that Syed and his allies are naturally eager to refute. In the second episode, for example, Saad claims that the fact that Adnan had other girlfriends and was a "player"—one with other objects for his romantic energies—proves he wasn't "obsessed" with Lee at the time of the murder. In that same episode, Koenig concedes that no one felt Adnan was suspicious, not "acting obsessed or menacing in any way." These characterizations suggest that the "obsession" being denied here is the mania of a misogynistic psychopath, a long-standing cultural trope. In the nineteenth century, psychiatrists who invented the notion of the insanity legal defense centered it around the concept of monomania, speaking of criminals who do not suffer from "lesions of the intellect" that lead to delirium and bizarre ideas but instead "lesions of the will" leading to moral turpitude.[29]

The second dimension of this moment has to do with how it is voiced. When Koenig says, "sad for sure, but not obsessed or anything" she is using a formal technique similar to what literary scholar Dorrit Cohn calls a "narrated monologue," a moment in which, while Koenig never stops speaking in third person vis-à-vis Syed, she takes on a voice as if speaking as Syed in his idiom and characteristic phrasing. Moments like this, Cohn theorizes, tend to commit the narrator to sympathy "precisely because they cast the language of a subjective mind in the grammar of objective narration."[30] But it also removes the denial of obsession from Syed's lips, nesting it in that same grammar.

Taken together, these two dimensions of the "obsession" statement mean that "The Alibi" lays across the map of its narration two entwined cliché concepts of obsessive thought, one that of the reporting detective "in too deep" and the other that of the monomaniacal killer maintaining a studied facade. And it sets these concepts into its world in the form of denials, both of which present as potential deceptions, in the first case by Koenig's use of a feigned aside that feels improvised, without truly attempting to fool the audience into believing that it really is, and in the second case by removing the claim from Syed's own voice, which does not appear in the podcast at all until a moment later. Using these moves, it is as if *Serial* wants us to keep chewing it over, the whole way through: Is obsession really "too strong a word" for either Koenig or Syed? Both? Neither?

Try as we might with these assertions of non-obsession, their sincerity never dispels their dubiousness and vice versa, producing a chain of deferral. This is what makes them available to iterative speculation, which becomes listeners' shift-work errand, by means of the show's episodic nature into distinct temporal periods. You can imagine a map of each episode that shows how the podcast leads you to believe and then disbelieve these twin denials of obsession, like a double helix stretching in ordered segments across the code of the series from week to week. The possibility of obsession is to *Serial* as the expectation of wonder is to *Radiolab*, in that it gives the show its signature emotion, its constant vibrating frequency.[31] So while the cultural style of public obsession with *Serial* was surely part of its media phenomenon—the convention of serialized releases of episodes week after week, whose delays provide time for these questions through fan engagement on social media platforms able to sustain and nourish ongoing reddits and subreddits, podcasts about podcasts—that style was not just the narcissistic projection of fans with the time and technology to grasp at straws thanks to a set of online social conventions formed around media affordances. In the piece itself, all the while, the unconvincing denial of obsession already twists like a worm.

Obsession and Genre

At this point I want to enlarge the framework and establish how obsession with obsession was not only in the language of this singular program but rather a fairly common feature that took several forms in late 2010s narrative audio, offering a way of mapping episodes, characters, and seasons that made aesthetic sense in a period in which communities surrounding the audio medium understood their field to be in a state of great transformation, a belief that even if overblown may also have been a self-fulfilling prophecy. Begin to search for it as a critic, and the obsession theme appears consistently across genres. In a 2015 release of an *Audiosmut* episode,

Radiotopia's *The Heart* told the story of a young man's attempt to satisfy a long-held sexual curiosity, describing an encounter in which the storyteller accepted a craigslist date with a kinky stranger who wanted to hire a partner to defecate on him. The anonymous speaker explains, in a statement that uncannily echoes *Serial*'s opening: "I know I wanted to do it because I have this deep-rooted interest or, I wouldn't call it obsession but . . . fascination with feces." Obsession is again a false parenthetical, a statement scripted and edited to be included in a way that mimics the form of an unscripted hesitation for dramatic effect.

There are also examples of hosts displacing their own obsession by putting it into the mouths of others. Consider the 2017 fiction podcast *Rabbits*, which begins as reporter Carly Parker sets out to learn what happened to her friend Yumiko, who seems to have disappeared into an alternative reality game. In her opening monologue, she reflects on the way she had neglected Yumiko and on the damage that the investigation has caused. "I had been working ridiculous hours, and when I wasn't working I was spending time with my boyfriend, who has since become my ex-boyfriend in no small part because of my focus on—or, what my ex referred to as my 'obsession' with—uncovering what had happened to Yumiko." Again, the term appears in an aside rather than in the stem of the sentence, ventriloquially displaced into someone else's observation. Obsession is often associated with loss or death. A fascinating case is "The Living Room," a haunting 2015 episode of *Love + Radio* that focuses on author and filmmaker Diane Weipert's *Rear Window*–like story of watching, for a long time, a young couple across the courtyard of her own apartment. Although it's clear she is obsessed with the couple (the first act of the story ends when she confesses that although she knew to look away, she instead reached for her binoculars), Weipert doesn't use the word "obsession" until it's clear that the man across the way has died of a terrible illness. Even then, the word is displaced into her account of her husband's observation of her observation: "I was in the kitchen and my husband called because he knew how obsessed I had gotten with this situation."

It's hard to imagine narrative moments more tonally, generically, and artistically different than these from *Serial*, *Audiosmut*, *Rabbits*, and *Love + Radio*. Yet they adopt nearly the exact same formula, and in some cases the very same words, to spotlight a vocalizing speaker who articulates obsession either by confessing it in the form of a denial that no one is supposed to accept or by remapping it into someone else's observation of themselves. Because this aesthetic pose winds itself into a question of trust, it is inevitably in crime stories that the obsession with obsession has the greatest force, and so it is impossible to extricate the style of these podcasts from the period of their rise. While notable crime and paranormal podcasts were produced early in the decade—*Generation Why* (2011), *Last Podcast on the Left* (2012)—the late 2010s true crime cycle began essentially in 2014 with the January debuts of the popular crime show *Sword and Scale*, later the subject of much controversy around the problematic behavior of its host, and acclaimed anthology *Criminal*, perhaps the most consistently accomplished show in the genre. The cycle then solidified that autumn with the first season of *Serial*. True crime swiftly rose to dominate iTunes charts and top-ten lists highlighted in *The Guardian*, *Vulture*, *Hot Pod*, and the *New Yorker* in 2015 and 2016. Until this point, most high-profile and mainstream podcasts were made by comedians who had moved into the medium to build audiences and extend their brands, so the arrival of *Serial* dramatically changed who could make podcasts and how, as productions streamed out of newspapers and magazines; cable TV networks; radio stations; publishers; public interest journalists; independent companies; new media entities like Wondery, Gimlet, and Pineapple Street; online venues like *Slate*'s Panoply network and the newly formed Radiotopia group; and subsidiaries of tech giants like Audible.

The crime genre was not all murder and mayhem. Several successful podcasts focused on corruption, like *Slow Burn* (the Nixon impeachment), *Crimetown* (Providence mayor Buddy Cianci), and *Ponzi Supernova* (Wall Street Ponzi artist Bernie Madoff), as others explored charismatic subcultures from self-help to cults, such as *Uncover: Escaping NXIVM*, *Oversight: Jonestown*, and *Guru*. A few were able to use crime to tell a deeper story in the context of an investigation—*Mogul: The Life and Death of Chris Lighty* is as much a history of hip-hop as it is a query into the music executive's suspicious death, as Nicholas Quah observed.[32] *Out of the Blue*, which focused on the University of Texas at Austin tower shooting, took the opportunity to collect oral histories of many who remembered that tragedy. Much of what was classified as true crime could also be thought of as difficult reflections on power, justice, and the material life of criminality. Several of *Love + Radio*'s most controversial and celebrated podcasts were about criminal lives: "The Wisdom of Jay Thunderbolt," about an unlicensed strip club run by a small-time hood in his house, and "A Red Dot," on the lives of former sex criminals. *The Heart*'s "No" series focuses on the complexities of communicating consent, revealing much about women and queer people's negative sexual experiences from a feminist point of view that previous eras of radio had altogether ignored. It is a great example of Michele Hilmes's point that podcast content lends itself to stories that are not just "intimate"—a cliché term for radiophonic communication—but can also be intensely *private*, relating "suppressed histories, dark secrets, terrible truths, not only on an individual but on a social scale."[33]

True crime shows were often investigations of how crime touched on political and social issues. Many of them investigated the history and social costs of the drug wars, from kingpin tales (*Chapo: Kingpin on Trial*) to stories about the rise of disproportionate incarceration of people of color as a result of racist drug statutes in the 1980s and beyond (*100:1 The Crack Legacy* and *The Uncertain Hour: Inside America's Drug War*), while others looked at social crimes through firsthand accounts of the people who lived through them, such as *The Home Babies*, a podcast about the mother and baby home scandal in Ireland, or *Making of a Massacre*, a 2018 investigation by ProPublica and Audible that dramatized a horrific incident in the Mexican border town of Allende perpetrated by the Zetas gang, with the bumbling involvement of the Drug Enforcement Administration. Many podcasts investigated the stories of those caught up in the carceral system, including the award-winning *Ear Hustle* about men in San Quentin and *Bird's Eye View* about women in Australia's Darwin Correctional Facility. Several podcasts emerged focusing on the perspective of those adjacent to the system, including spouses and families of the incarcerated, such as Falling Tree's *Prison Bag*, and Charlotte Rouault's *Under the Other Pain*, while the podcast *70 Million* looked at the way prisons impact their surrounding communities and how the millions of Americans with criminal records negotiate their lives on the outside. *Criminal* continued throughout the 2010s to be the most innovative and philosophical crime anthology of its era, taking issues from the life of a prolific streaker, to the crime trial of one of the last living Nazis, to the fallout of historical off-book adoptions, to the idiotic clerical police errors that led to the shooting of an innocent Black man driving at night.

That said, it remains true that these often difficult explorations arose against a relentless stream of pulp murder podcasts, many of which focused on reexamining grisly cases that had been generating profits for sensationalist film and TV for decades and brought little in terms of social, ethical, or political challenge to their subjects: *Hollywood and Crime* (the Black Dahlia), *Inside Psycho* (serial killer Ed Gein), *Stranglers* (the Boston Strangler), and *The Vanished* (an anthology of many cases). These shows all came from Wondery, which had a strong vertical in the true crime space, finding particular success with tales of predatory male rogues—*Dirty John*, *Dr. Death*, *The Shrink Next Door*—that would go on to become TV series and films. Wondery's brand strength shaped listening habits. In 2021, Pacific Research's Dan Misener did an experiment in which he downloaded a list of the top four hundred podcasts on Apple Podcasts and for each one tracked its "listeners also subscribed to" list, thereby generating a rough map of listening "neighborhoods," at least as the algorithm apparently understood them.[34] Figure 2 is a modified version of Misener's visual map of these neighborhoods; in it we can see "mainstream programs" such as the *Joe Rogan Experience* and *Serial* (see fig. 2 rectangle area) that link to many other areas, including a neighborhood for NPR and newspaper-based journalism podcasts, another for money and economics podcasts, as well as distinct enclaves for children's programming and right-wing talk programming. Most remarkable is the fact that there are *two whole neighborhoods* for true crime (circles), one of them almost entirely formed out of Wondery content.

Of course, the popularity of the form can't be attributed to one show or one company, but the tide from this productive area carried many other projects along with it. Independent Hollywood history podcaster Karina Longworth found so much success with her series on the Manson murders on *You Must Remember This* that it was spun off and repackaged as an independent podcast around the time of Charles Manson's death in 2017. Detroit Channel 4's *Shattered* called attention to the disappearance of Jimmy Hoffa, the Oakland Child Killer, and famed police informant White Boy Rick, while Andrew Jenks's *What Really Happened* focused on the trial of Jussie Smollett, the tawdry sex conviction of NFL owner Robert Kraft, and the monstrous pedophile crimes of Jeffrey Epstein. Erstwhile filmmaker Payne Lindsey found success with the *Up and Vanished* podcast about disappeared high school teacher Tara Grinstead, which spawned many seasons about other disappeared women, as well as podcasts dedicated to better-known killers, such as *Monster* (the Zodiac Killer) and *Atlanta Monster* (Atlanta child murders). There were also original pieces about locally famous cases that required exhaustive reporting, especially when they stressed a sense of locale and civic corruption, including *Thunder Bay*, *Breakdown*, and *Murderville*. *West Cork*, a 2018 podcast by Jennifer Forde and Sam Bungey about the infamous murder case of Sophie Toscan du Plantier in the titular Irish county, took years to investigate and produce; once widely released it became Audible's single most listened to podcast. Thanks to the precedent set by *Serial*, such works were greenlit even by budget-conscious public radio heads, and many productions

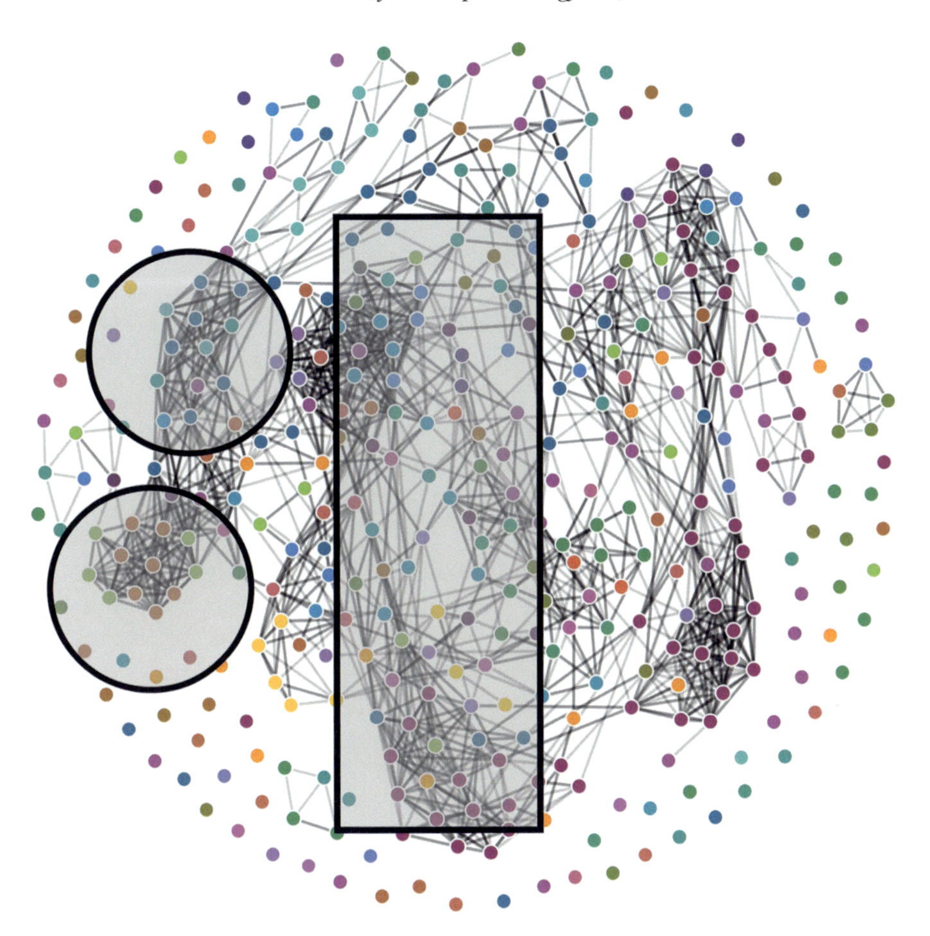

Figure 2. Podcast listening neighborhoods. Each dot represents a popular podcast, and each line how the Apple algorithm recommends other podcasts to that podcast's listener. The circles identify "true crime neighborhoods" and the rectangle shows the mainstream of the most popular podcasts. Image courtesy of Pacific Content.

became Peabody and Third Coast award winners thanks to an emphasis on broader social ramifications, such as Minnesota Public Radio's *74 Seconds*, on the trial of the police killer of Philando Castile. In this way, vital works about social and racial justice found expression thanks in part to a market for sensationalism that *Serial* had galvanized and that pulpy shows like *The Disappeared*, *Small Town Dicks*, and *Casefile True Crime* nourished.

Mapping the Ceremonial Confession

In many of these podcasts, the language of "obsession" appears at the start of an episode, season, or series, often to justify the existence of the podcast and to humanize both victims and reporters, yet the language of obsession is also often framed as if there was something immoral, embarrassing, or naughty about using it. Obsessions are, for this reason, frequently disclosed apologetically. *Accused*, a 2016 podcast by Amber Hunt and Amanda Rossmann of the *Cincinnati Enquirer*, begins with a reflection by Hunt about her obsession with the case of Elizabeth Andes, a twenty-three-year-old woman murdered in 1978. "In some ways I feel like a stalker," she says, tracking Andes's every waking moment and conversation in her last weeks; she keeps Andes's picture on her wall, on her desktop, on her phone. Obsession could be the prompt for a relationship around which a whole program is launched, such as *My Favorite Murder*, a podcast that began in 2016 premised on how a mutual obsession with murder stories brought hosts Karen Kilgariff and Georgia Hardstark together, forming a community of "murderinos" with a similar obsession. This was also true for limited series shows with tantalizing premises. Pineapple Street's 2020 podcast *Winds of Change* centers on a thirdhand story about how a famous Scorpions rock song about perestroika was actually written by the CIA, a story that reporter Patrick Radden Keefe and his friend Michael had been "utterly obsessed with" for a decade, something they describe as "a fixation," a story "so compelling" they couldn't give it up. In other cases, an excess of obsession could be blamed for a program's downfall. The most infamous case of this phenomenon from the late 2010s was Rukmini Callimachi's podcast *Caliphate* for the *New York Times*, a Peabody Award winner that focused on detailed interviews with Canadian Shehroze Chaudhry, who claimed to have inside experience of what life was like as a recruit for the Islamic State in Syria and Iraq but turned out to be a fabulist. *Times* executive editor Dean Baquet, in his official retraction of the show, blamed his team's attachments and emotions, conceding in one news story that his fault was that "we fell in love" with the story.[35] It's an odd excuse, unless viewed in the context of a listening culture that took obsessive thinking so much for granted that it is normal and even acceptable to think about a journalistic stumble in the same terms as a jejune infatuation.

An interesting case is *Missing Richard Simmons*, a 2017 top-of-the-charts podcast that belonged to a genre of celebrity story pieces (*Dolly Parton's America* and *Making Oprah* were great examples in this period) about the titular fitness guru, who suddenly retreated from public life the previous year, no longer returning phone calls from friends, leading fitness classes, or making public appearances. In the podcast, filmmaker Dan Taberski, a friend and student of Simmons, begins like many podcasters with an affective claim that despite having no particular fitness interest, he was "instantly and completely all about Richard" the moment he was invited into his world, meeting his dogs and seeing his famous doll collection. When the gregarious Simmons went missing from public life without warning not long after, Taberski grew worried about him. Over several episodes, Taberski tells the story of his subject's fame and life, while exploring theories ranging from a knee injury and the death of a beloved dog to the machinations of a manipulative maid. By the fourth episode, the whole project begins to feel forced; when a phone call emerges from Simmons himself explicitly claiming he isn't being held hostage by anyone, Taberski takes it as possible evidence that maybe he is *indeed* being held hostage.

The podcast was criticized by a number of authors as an undue invasion of privacy—*Mashable* answered the question of "why everyone's so obsessed with *Missing Richard Simmons*" with the succinct theory that "we are so nosy."[36] Separate from the question of whether it fails as a piece of ethical journalism, I'd argue that as an aesthetic matter it is a failure of successfully matching the scale of Taberski's obsession to that of his subject's circumstance. The narrator's map of his own fascination does not correspond to the map of the conundrum he faces. Many podcasts begin with a narrator in a state of bashful embarrassment about their obsession, something they need to divulge and then overcome, ritually. It is a kind of ceremony. As we begin to hear the story through the narrator's ears, we gradually forgive the sin of obsession, as we are persuaded to believe how justified the feeling is after all, becoming obsessed ourselves. In this way, a host's sheepish moral concession of fault opens the door to a normalization of the impulse behind that fault. *Missing Richard Simmons* is not a long podcast, yet it grows awkward in part because it breaks both sides of this ritual code, neither expressing customary mortification about obsession at the outset nor successfully recruiting us into that obsession over time.

Perhaps *Missing Richard Simmons* might have taken a different route, one that many podcasters did in this period when they deliberately engaged with their own obsessive tendencies in a self-critical manner. In *Another Dead Man Walking*, for instance, a 2015 Sky News podcast about a death row inmate, reporter Tim Woods explains he became involved in the case simply because of his longtime "fascination" with the death penalty, accepting the role partly out of a "dark sense of exhilaration." Obsessions could also be presented with deep contrition. In a searing two-part 2018 pod about media panics in *Revisionist History*, Malcolm Gladwell follows up on the story of Margit Hamosh, a persecuted lab scientist he'd encountered in his days as a *Washington Post* writer chasing science fraud back in the 1990s, a story that shows the great danger of reportorial drive to break stories about infinitesimal grant-writing errors. In the first episode, he uses the term "obsession" each time he indicts his own former practice, his voice trembling in bitterness. In these examples, obsession is presented as an undisguised fissure in the edifice of reliability, the concomitant curse of a journalist's appetite for knowledge. In this way, the podcast hosts a well-worn melodrama of professional journalistic ethics and the temptation toward its subversion through an impulse to fall too in love with a story (as in the *Caliphate* case). The presenter is in a moral struggle with themselves: they appear in a state of temptation by their own impulses that they first shamefully feed before our ears and then overcome. It is this deep implicit drama that can lend a current of morality fable to the obsession narrative, a subterranean language that ratifies the position of the storyteller with a lure of redemption, the thing they are really chasing through the episodes.

Mimetic Obsession

The obsession with obsession is, as I hope is clear by now, a guiding passion in how podcasts in this period were written, structured, and produced. It is a safe bet that if, at the start of a show, an obsession is introduced by a narrator as a justification for the very existence of the podcast, the thing that willed it from nothingness into existence, then this obsession is likely to be instantly remapped from the center of the project to its periphery by means of displacement, denial, confession, or some other mechanism of banishment. From this periphery, the obsession will remain all the more motivating, establishing the show's very perimeter and duration—and, it is to be hoped, the obsession will become mappable into larger conversations in social media among fans who build an affective attachment with the podcast. Obsession isn't just an empty feature or theme; it makes things happen, thanks to a quality of a suspicious beckoning, whether it calls a podcaster to a story or a fan to a forum. And there is another way in which obsession mapping resembles desire, to be found in a subgenre in which obsession is shifted from the main enunciative "host voice" of the show into a surrogate responsible for bearing that obsession, with the effect of neutralizing any latent moral heaviness and signaling a tonally lighter story. The figure of the "surrogate obsessive" is, for this reason, a key trope

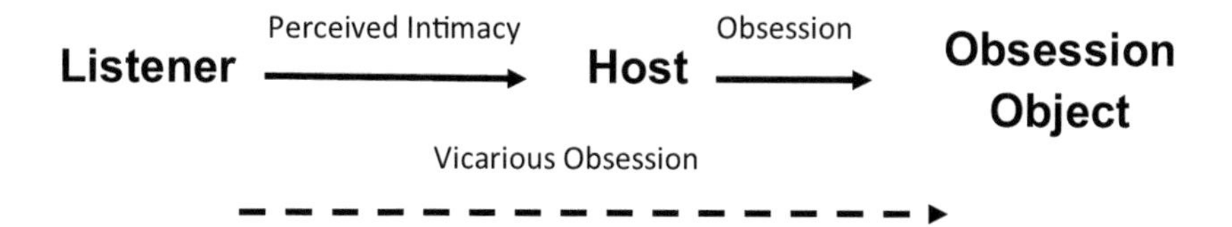

Figure 3. A typical obsession podcast, with a listener linked to a host through a perceived intimacy, a host linked to an object through obsession, and a listener therefore linked to the object of obsession vicariously. Diagram by the author.

in late 2010s narrative podcasting, one that has a reflexive quality, as the relationship of the host to the surrogate mirrors or recapitulates that of the listener to the host. What is formed in many cases is what we might call a triangle of mimetic obsession.

To explain, let me back up for a moment. In a typical "obsession podcast," such as the ones discussed above, a Listener's vicarious interest in an Obsession Object is usually mediated by an intimately accessed Host, who is hopefully able to enlist the Listener into the obsession wholeheartedly and even get them to pass it along to others in a fan community. That mapping looks something like figure 3, where each line represents a kind of excited desire.

The "intimacy" in this scenario is in my view less an inherent property of the medium and more a result of a few controlled aesthetic conventions that are so commonplace in radio as to go unnoticed. Some of these include close miking of most vocalizing subjects with low direct-to-reflection sound ratio; long-form narration; the use of a higher gain for hosts than for other voices and sounds in the space of depiction, creating a contrast of spatial alignment; and even the very fact that we typically have one main host, who tells us their first name early on.[37] Those are some features of the "aesthetics of intimacy" you can find in the production setting in a contemporary podcast.[38] In the reception setting, meanwhile, "intimacy" often involves the element of touch, be it with earbuds wedged into our ear canals or with the all-encompassing embrace of noise-canceling headphones. We also often listen through a smart phone or other device that is "intimate" to us, both in the sense of being our personal item and also in a sense as representing a fulcrum between our place in the physical world and our place in the digital sphere. Indeed, among early theories of what made podcasting special, the conjunction of mobility and privacy, promising what Mack Hagood calls "orphic listening," was seen by many to be among the chief attractions of the medium.[39] It was around the time of *Serial*'s release that this mode emerged into prominence, too—in 2013 most podcasts were downloaded to personal computers, but in 2014 and 2015 data suggest that tablets, MP3 players, and smartphones received the majority of RSS downloads from Libsyn, the largest syndicator at that time.[40] These are some of the aesthetic and technical ways in which radio's "inherent" intimacy is actually aesthetically and technologically formed. Only afterward is this system made to seem universal and inherent to the medium. Many say that an intimate, richly produced voice touching our ear and guiding us into a story is the "essence of radio," but this belief is only an ideological reification of the system described above; it is equally fair to say that radio's "essence" is to be found in static created by distant, indecipherable space weather, which sounds about as dis-intimate and alien as it gets.

So that is a typical or "linear" mapping of an obsession podcast. In a triangular or "surrogate" mapping structure, by contrast, the Listener grows mimetically interested in an Obsession Object only because they feel intimately connected to a Host, who expresses a vicarious interest in the Object with which their chosen Surrogate is truly and deeply obsessed. Figure 4 shows the general contours of this situation.

Obsession requires, in these cases, a two-stage set of transpositions in which the listener feels a kind of excitement because their interest is mediated by the excitement of two other people. In both linear and triangular configurations, as in René Girard's well-known framework of mimetic desire, the particular object of the obsession itself may in fact be outside the listener's interests and may even be unpleasant to their sensibility, yet by being mediated by another individual's interest, initial disinterest or even outright antipathy becomes fascination.[41] And when surrogate obsessives are used most effectively, it is the surrogate's interest itself, as ratified by the intimate host, that captures the interest of the listener, not the particular material object to which that interest is oriented, the attainment of which is nearly immaterial to the pleasure of the piece.

The first episode of *Criminal* contains a good example of this phenomenon. Focused on the "staircase" murder of Kathleen Peterson in Durham, North Carolina, which had already been the subject a famous film about the trial of her husband for the murder, the podcast looked at an outlandish theory that an owl strike led to Peterson's perplexingly bloody death, a theory touted in the podcast by Larry Pollard, one of Peterson's neighbors. Podcast host Phoebe Judge explains by way of introduction: "Remem-

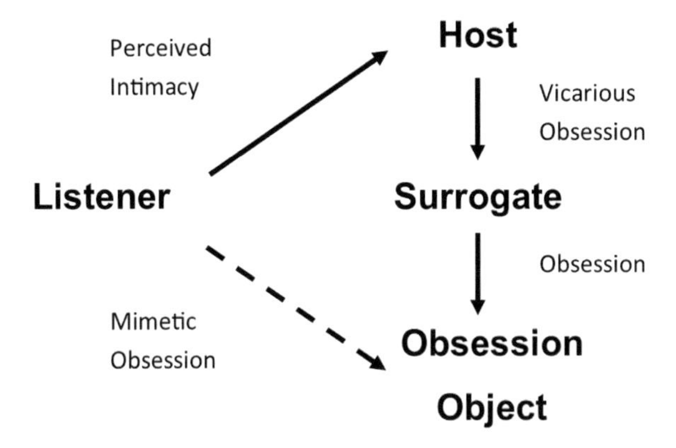

Figure 4. A surrogate obsession podcast showing how a listener feels intimacy with the host, who vicariously experiences obsession with an obsession object through a surrogate. This double mediation produces what I call "mimetic obsession" in the listener. Diagram by the author.

ber, Larry Pollard isn't a scientist, he isn't a detective, he isn't even an owl expert. He's just a lawyer. Not even Michael Peterson's lawyer. He built an elaborate theory around microscopic owl feathers, and he's become kind of obsessed with it." It is curious that Pollard's lack of professional investment needs expression here (this construction resembles very closely a line in *Serial*'s opening, by the way, where Koenig says, "I'm not a detective or a private investigator. I'm not even a crime reporter.") before he can qualify as a true obsessive, a possessor of theories that qualify as "elaborate." But the point is that few listeners will start out the podcast as owl strike hypothesis afficionados, and few will leave it that way, yet by mediating Pollard's obsession through Judge's interest in it, a temporary and enjoyable form of mimetic obsession is offered to the listener, one that has the feeling of control that monomania can afford, without its debilitating pathological properties. In other cases, podcasters meet with amateur detectives who are unrelated to the crime but who had been looking into specific cases for years as a hobby. These figures often become colorful characters in these shows, painted as highly emotional, with unabashed passion counterbalancing the podcaster's professed cool reserve. In *Bear Brook*, for example, reporter Jason Moon meets up with amateur investigator Ronda Randall, a self-professed obsessive (she describes herself as "dogged," "tenacious," "a pit bull") who has for years been combing for clues about a horrifying murder involving two unidentified bodies found in a barrel by wandering kids in the titular nature preserve in New Hampshire.

In both cases, the fact that the listener's intimate hosts (Judge and Moon) find the obsessions of the surrogates (Pollard and Randall, respectively) so compelling generates a powerful rhetorical impulse for the listener to do so as well. If our trusted and admired host, a being that "intimacy" has aligned us with, is so interested in this person who has an idea about some whatsit, then we feel we should be too. It is as if the residue of the vicarious interests of our hosts sticks as we pass through them and their surrogate toward the set of facts behind these mediating presences. This can even be the case when the enthusiasm of the host is underplayed. In these years, one of the most requested stories on the prolific anthology *Snap Judgment* from WNYC, "06 Female," told the story of a legendary alpha female wolf of the Lamar Canyon Pack in Yellowstone National Park. The surrogate obsessives, biologist Doug Smith and technician Rick McIntyre, trade well-worn tall tales of the animal, known for her beauty, intelligence, and physical strength, the colorful western heroine beloved for her harrowing escapes from rival packs, ability to take down a bull elk singlehandedly, and daring avoidance of radio tracking darts. When the wolf was shot by a hunter in 2012, the men entered a kind of grief, one that we feel as well. Through the entire episode, however, host Joe Rosenberg guides the story with a very low, understated vocal style, showing nothing like the enthusiasm of his surrogate narrators. Nevertheless, by dint of this framing presence, the narrative generates much the same affect. The mediating presence need not contribute to the affective energy of a story; it need only amplify it.

In storytelling podcasts, the use of surrogate obsessives could also create an effect of generosity. Contrast the seriousness of Malcolm Gladwell's interrogation of his own former obsessiveness described in the previous section with the kind of obsessions we find in *Heavyweight*, Jonathan Goldstein's podcast in which he helps friends and acquaintances find the moments their lives went wrong and deal with regret and forgiveness, a show that probably embodies White Gen X middle age better than anything else in the podcast sphere. In one memorable episode, Goldstein's friend Gregor mourns the loss of meaning in his middle years by seeking out self-worth and recognition by way of a quest to reacquire a set of CDs that he once lent to music superstar Moby and that formed the basis for one of the latter's hit records. The "quest of a friend" narrative lends itself very well to obsessive surrogacies. In "Punks," an episode of the WNYC show *Nancy*, hosts Kathy Tu and Tobin Low help their friend Kai doggedly track down a long-lost film he had seen in 2000 called *Punks*, a campy beloved exploration of urban Black queer life that had once been common at gay film festivals, in the hopes of showing it to a younger queer-of-color friend just coming out of the closet. Tu and Low helping Kai help his friend creates an interesting effect—the episode isn't just about *making* mutually supporting outsider community; it *is* a mutually supporting community, and the episode ends in a screening expressively embodying this.

Mapping obsession into friendship is in many ways the whole idea behind *Mystery Show*, an unfortunately short-lived 2015 hit for the Gimlet network. In the first episode of the series, host Starlee Kine relates a childhood story of finding a mysterious old freestanding safe in her grandparents' garage with her little sister. After long speculation, they eventually discover the code and breathlessly open the safe, only to find it empty. "I've been obsessed with mysteries ever since," she says, wrapping up her origin story. Rather than pursuing her own fascinations, however, Kine presents herself as a kind of detective figure, unraveling the mysteries of others, like her friend Laura, who once rented the film *Must Love Dogs* from a video store but then couldn't return it because the store suddenly vanished the next day, or her friend David, obsessed with discovering the "maximum morning height" of actor Jake Gyllenhaal, a matter of acrimonious faux controversy in online forums. *Mystery Show* makes high comedy of the trivial mysteries that friend communities exchange to amuse one another. It is interesting that although the scale of detail that both Malcolm Gladwell and Starlee Kine investigate is identical—the infinitesimal detail of a grammatical construction in a government report, the inconsequential measurement of an actor's height—Gladwell's high dudgeon couldn't be more different from Kine's laughter as her friends relate their pet mysteries.

A curious red thread linking *Heavyweight*, *Nancy*, and *Mystery Show* is that all three have tangible media as their quest item—films, CDs, and video tapes. The map of surrogate obsession leads to an element of playable media for the podcast itself to hopefully exhibit; the thing sought for a friend, or for a friend of a friend, is also a thing sought for the final edit and mix. In this way, the quest for emotional closure is paired with the quest for production completion. Considering this trio of shows together also inevitably raises an implicit issue about positionality that is clearly present but unstated in the podcast. To what extent does the mimetic obsession structure require or form an alignment of identity between listener, host, and surrogate? All three of these podcasts are directed toward middle-class listeners at or nearing middle age, much like the surrogates chosen in all three cases, but one chain of mimetic obsession passes through a White cis woman host, another through a White cis man host, and a third through a queer-of-color framework. Since listeners necessarily bring their own identities to the triangle of mimetic obsession, a future area of research might lie in the heterogeneous ways in which surrogacy is experienced by listeners based on the positionality of the individuals in the triangle. The three cases also show how the stakes of the object can vary. The *Nancy* obsession object has a lot of importance, as a part of a rite of passage and community memory, while *Heavyweight*'s object of obsession is important only to a particular person, as a fetish object that encapsulates Gregor's ennui. *Mystery Show*'s objects of obsession are low stakes by design; no lives hang in the balance in discovering Jake Gyllenhaal's maximum morning height.

Perhaps no podcast of the 2010s profits off the sheer esotericism of the obsession object more effectively than *99% Invisible*, given its focus on design elements that, by the very mission of the program, are supposed to be unnoticed, which means that people who devote their time to understanding these elements come across as lone eccentrics or loveable cranks. In these years, this flagship show of the Radiotopia network would feature interviews with people obsessed with Hong Kong's Kowloon Walled City, with buildings that used to be Pizza Huts, with the history and use of sand in modern building materials. A rather beautiful case is "The Accidental Music of Imperfect Escalators," an episode in which host Roman Mars introduces then intern Sam Greenspan, who in turn introduces Chris Richards, a pop music critic fascinated by how the imperfections of wear and tear gave each Metro stop escalator in Washington, DC, its own specific sound. The remainder of the episode explores Richards's obsession with the "mating whale" sounds of the Farragut North escalator, the "Indian drone" music at U Street, and the "aviary of chrome throated ravens" at Columbia Heights, sounds that come across as bearers of secret musical wisdom. Mars's excitement about Greenspan's excitement about Richards's obsession resembles the hall-of-mirrors quality in the "Obsessed with *Serial*" video cited at the outset of this chapter. In *Mystery Show* the magic trick of mimetic obsession fools the listener into playing along with the play of the obsession by way of imitation, but the *99% Invisible* episode does something even more remarkable, teaching us to hear these strange sounds as Mars hears them through Greenspan's ears through Richards's ears.

From Obsession to Commitment

An extreme case of mimetic obsession can be found in *The Butterfly Effect*, one of a cluster of programs created during a brief production cycle associated with Audible's paywalled Channels service in the late 2010s. In seven episodes, host Jon Ronson takes a gonzo journey through a series of repercussions that flow from the advent of free pornography on the internet, which was the brainchild of entrepreneur and data expert Fabian Thylmann of PornHub. We learn of the drudgery of adult film actresses and actors driven to work overtime to create content, of a pastor's daughter confessing to porn addiction, of out-of-work porn actors shunned by potential alternative employers and even by their own children. Perhaps the most compelling storyline in the series comes in the second episode: a set of interviews with Dan and Rhiannon, veteran porn producers in the San Fernando Valley who try to make up revenue lost to streaming aggregators by creating "customs," bespoke short films for anonymous fetishist clients that range from re-creations of lifeguard TV shows to elaborate fantasies in which performers pie themselves in the face and wrestle blankets. Ronson paints a picture of a hidden industry based on obsessions as an emergent form of post-PornHub labor. He himself soon becomes obsessed with an urban legend in the custom porn production community about an enigmatic foreign postage stamp collector who hires actresses in the Valley to burn his priceless stamp collection on tape. Over the course of several episodes, Ronson manages to contact "stamps man" and learns the origins of his picturesque philatelic fetish. Oddly, the pleasure with which Ronson ferrets out the man's story feels far more uncomfortable than the fetish itself.

At any rate, the series insists on a fascinating generosity that porn producers offer their clients, treating their fragile needs gingerly, aware that many come from a place of humiliation. In the concluding episode of the series, custom porn producers Dan and Rhiannon receive one of their most unusual requests. A customer asks for a video of a woman sitting in a room on the floor, cross-legged, talking to the camera to convince him not to commit suicide. In a panic about the client's safety, Dan and Rhiannon scramble to arrange a shoot with a performer named Riley, free of charge. Ronson calls it a leap of faith, hoping that the video will "be a flap of a butterfly's wings for [the customer], then everything will start to get better." We hear Riley address the invisible customer—importantly, as if it were addressed to us—and counsel him to remember the beauty in the world. Ronson reflects, "The more time I spend in the world of custom porn, the more I can see how much the producers and their clients have in common. They're all working out their issues together by producing these sweet, strange films." The map of surrogate obsession ends here in the utopia of anonymous, hidden community therapy.

Whether the thing being obsessed over has to do with murder, a reclusive fitness guru, the height of a Hollywood heartthrob, or a philatelo-pyromaniac fetish, a great many narrative podcasts are structurally pre-occupied by the enchainment of obsession between narrators, surrogates, and featured speakers. Obsession seems to abhor straightforwardness and is therefore "mapped" in one way or another, be it into a second voice, into a second or third person, across a set of episodes, between narrator and theme, among podcasts in a network, or across social media through fans. It is a hot potato passed between a speaker and another, a thing retreated from apologetically, or a way of forming parasocial communities.

Although my brief here is to focus on how obsessions are performed and poetically structured, it's important to recognize that many of them were not *just* performances. The people these podcasts are about are often people in states of deprivation and injustice, for whom a quest is not only "obsession"—the surplus of affect lending color to a drab life or suturing a fear of the outside—but also a righteous response to a form of structural violence actively being done to them. The obsession podcast was, for all its salaciousness, a way to expose that violence and thereby represented a new level of mission and sincerity entering the podcasting space after a decade of mainly comedy and chat programming. In growing the industry and achieving a new aesthetic proper to it, obsession podcasting could convey the impression that it "unlocked" public radio by allowing many producers, often in maddeningly underfunded news outlets and radio stations, to execute projects that had long been a simmering wish but had been held back by funding constraints and other gatekeepers. For a decade before *Serial* there had been "nerdcast" talk-based podcasts that allowed best-friend hosts to discuss the things they secretly collected or subcultures they belonged to. To this body of work, public radio–trained narrative podcasters brought their own hobbyhorses, often with aims of solving pressing social problems and shedding light on contexts of relentless racial violence, festering structural failures in criminal justice, or stories that affected them personally.

In the fourth episode of 2017's *Heaven's Gate*, for instance, Pineapple Street Media producer Jenna Weiss-Berman takes time out from the show's history of the infamous titular cult to interview the narrator of the show itself, *Snap Judgment*'s longtime host Glynn Washington, who grew up in a similar cult and speaks to the elements of the story in a personal way. In 2020's *Mississippi Goddamn*, *Reveal*'s Al Letson traveled back to the town of Lucedale, Mississippi, where he had done unrelated reporting ten years earlier during the time of the 2010 Deepwater Horizon oil spill, to fulfill his promise to the family of a dead young Black football player, Billey Joe Johnson Jr., who died during a suspicious police arrest. In 2021's *Suave*, a Pulitzer Prize winner for Futuro Studios, *Latino USA* host and NPR reporter Maria Hinojosa told the story of a man jailed for the rest of his natural life for a terrible crime he committed as a juvenile, a subject she had been calling and interviewing since 1993, as a new court ruling brought about a sudden change of his circumstances.

Like the first season of *Serial*, these are passion projects by longtime public radio figures—in these cases, importantly, all reporters of color—who are as much subjects as they are storytellers in the long-form context. Unlike *Serial*, however, all three of these producers tell their stories in order to fulfill decades-long and in some cases lifelong commitments to the people in their stories. As a result, these podcasts feel less like indulged pathological fascinations and more like virtue narratives about reporters keeping their word. This change coincided with a demographic shift in activist podcasters, with an increasing number of racial and ethnic minorities, LGBTQIA, women, and others telling stories related to their identity group (although as one analysis showed, this inclusivity did not extend to the working class).[42] As the decade waned and the vogue of performances of obsession began to dissolve, it made way for something else that might replace it, like the inevitable antipode of a dialectical structure sliding into place. From an aesthetics of obsession that had grown unstable came an ethics of commitment focused on "for whomness" and "about whomness," a topic to which I return in chapter 2.

Obsession as Critical Strategy

Before moving past the historical account that this chapter has offered of the rise and fall of obsession as an aesthetic agenda in narrative podcasting, let us first move orthogonally out of that account and ask: What should we "do" with obsession as a critical and theoretical matter? So far, my approach has implicitly been to unpack, historicize, categorize, and demystify it, to bring a kind of discipline to it by way of a broad taxonomy that imposes a grid across a great many podcasts. Such a criticism looks at obsession as a thing to be confronted, a thing distant from the critical analytical position, as if criticism itself were an enterprise immune to the temptation toward obsessive thought. Nothing, of course, could be further from the truth. It was precisely out of the development of an obsessive relation to podcasts that a great many scholars in the 2010s elevated podcast stories to an object worthy of study. With that in mind, it has come time to return to the first season of *Serial* and fold the argument about obsession back on itself once more, asking what it would mean to "do" close podcast reading in an obsessive manner rather than attempting detached observation, as I have been doing above.

Here I take a cue from Rita Felski's recent challenges that rethink the tradition of criticism that is rooted in demystification and dislocation, what she calls the "fear of stickiness," foregrounding instead the richness that textual attachments, attunements, and identifications offer. A stress on responses to our own attachments—especially when these attachments are vexed and complicated—seeks on the one hand to take seriously the enervations that art awakens on the critical sensibility, while on the other hand keeps the actants involved in generating artwork in view rather than subsuming them behind intensities of sensation. Obsessive listening can offer a way into critical engagement with a podcast that looks for elements of a creative team's narrative approach and at the same time listens for elements that shape our own affective attachments to a piece. As an experiment in that idea, I'd like to return to where the chapter began, with a reading that inverts the emphasis on characters and context that normally engage *Serial* listeners and begins instead by connecting two cues in its audio poetics, using these cues to generate a monomaniacal reading strategy around a "sticky" sonic feature that can be meticulously tracked across the landscape of the season, yielding a richer reading of the piece's subtexts than I think has been hitherto available.

The first cue I want to highlight is *Serial*'s reflexive fascination with specificities, often horrible details dealing with suffering, death, and abjection. The podcast goes into great detail, for example, about items found at the scene of Hae Min Lee's murder, including an unused condom, a liquor bottle, a rope, shell casings, bullets from two different guns, and Blockbuster video cases, despite the fact that none of these specificities amount to a clue that comes up again in the podcast. Details are also foregrounded that have no connection to the case at all yet amount to passive, cautious information, like the fact that Syed was proud of being able to prepare omelets with caramelized apples and onions for his friends in prison or, in the second season, the Taliban's particular fondness for Mountain Dew. Sometimes details are exhausting, particularly the accounts of how cell phone pings can identify individual cell locations or the vagaries about whether there was a payphone at a long-gone Best Buy electronics store in the first season. Episodes 5, 6, and 12 focus on the single piece of evidence that looks worst for Syed, a call from his cell phone to his friend Nisha, which puts him in accuser Jay's timeline; in the second season the question of whether AWOL soldier Bowe Bergdahl was taken to Kuchi nomads prior to being turned over to the Taliban is considered in episode 2 and again in episode 7.

The second cue I'm following is how the show prides itself on a sophisticated understanding of the voice as a nonverbal communicative instrument. In the second season, for instance, Koenig's narration not only focuses on what her interviewees say about the role of the Haqqani network in the politics of the AfPak region but also notes the emotional weight of the pause that always seems to precede any explanation of that role. In the first season, when Koenig finally gets hold of potential exculpatory witness Asia McClain on the phone, she explains the depth of meaning in McClain's sigh on the phone line when she realizes what her testimony could mean for Syed's exoneration. In the third season, the show often uses characterizations of the voices of interviewees to give them shape. Koenig remarks, for example, that accused murderer Davon Holmes speaks in repetitions when things stick in his mind, as if "skipping a mental stone."

The obsessive listening strategy I want to pursue joins together these two *Serial* cues by mapping a minor nonverbal pitch-shifting detail in Syed's voice that appears frequently when he speaks. I came to my interest in Syed's speech patterns in part because, like him, I am a child of South Asian immigrants and am perhaps more attuned to how second-generation young people use language as a way to fit into a North American context and relate to other people of color from other backgrounds, an interesting example of Nina Eidsheim's insight that the voice is a manifestation not of an identity but of a set of shared practices that are in a state of constant creation, based on context.[43] Syed's use of slang turns of phrase and excessively colloquial intonation was "sticky" to my ear, and I became interested in how a particular recurring detail in Syed's voice interrupted this feeling. To visualize the detail, I used a pitch-tracking program called Drift, developed by programmer Robert Ochshorn in part with help from a grant that the poet and scholar Marit MacArthur and I won from the National Endowment for the Humanities specifically to work on measurement tools for performed texts. Drift uses a textual alignment tool called Gentle to match a transcript of performed speech with its corresponding audio file, mapping the audio to the text, and then generates measures that give qualitative understanding of some of the elements of that performance; this program has been successfully used to analyze audiobook speech, poetry readings, and the like to show how readers modulate pitch, speed up or slow down over the course of a line, or obey or disobey caesurae.[44] Using it, you can discern, for example, whether a speaker accused of being "monotone" or using "uptalk" is exhibiting these features in a measurable way or if these accusations might have an ideological root we can unpack.

Figure 5 is generated using these tools, to examine one of the first lines that Syed utters in the series—"it didn't exist in me, you know what I mean?" referring to any hate toward Hae Min Lee—through the pitch tracker, whose values you can see expressed in a logarithmic scale on the x-axis.

Notice the jump between "didn't" and "exist," a sudden shift in pitch from approximately 131 hertz (which is the vicinity of Syed's mean pitch of 146 hertz throughout his first passage of speech) to 349 hertz and then a shift back down again. It is not an isolated deviation. In the first two hundred words that Syed speaks in his first appearance in the episode, there are five instances of this very same brief upward vocal shift of about 190–220 hertz as visualized on Drift's logarithmic scale: "I had no ill **will** toward her. [. . .] It didn't **exist** in me. [. . .] No one's ever been able to **prove** it. [. . .] The only thing I can ever **say** is, man I had no reason to **kill** her." Later on in the episode, Syed's voice does it again in his recounting of what happened the day of the murder when he says "**you** know," a phrase that is Syed's most common pause filler, then again when he seems to throw up his hands in frustration toward the end in "perhaps I'll **never** be able to explain it." When he learns that there is evidence that might convince Koenig about his innocence but does not really help his case near the end of the episode, he says that "on a **per**sonal level" he found it rewarding.

This pitch-shifting detail of Syed's speech, which does not appear to be an intentional emphasis, is what we colloquially call a vocal "crack," a temporary loss of control producing a shift from the "chest voice" to the "head voice" with the effect of a falsetto. In music, as Martha Feldman has pointed out, the vocal crack takes place when a movement of the larynx exposes an aspect of the mechanism that should remain hidden in the vocal-

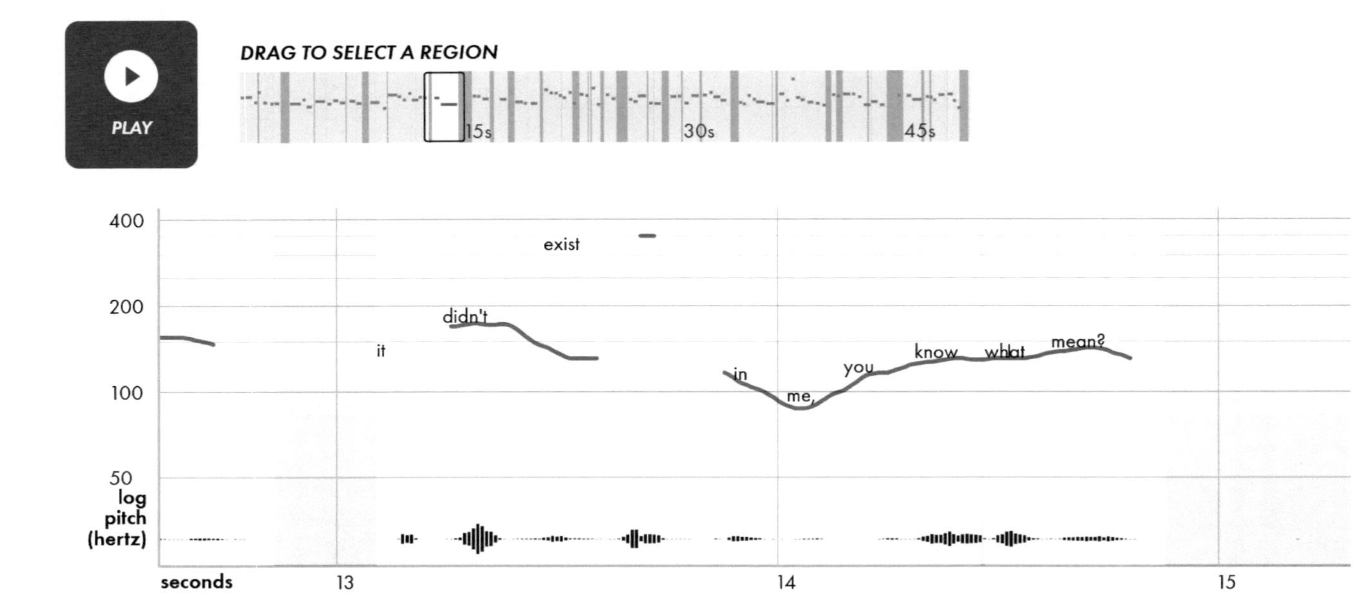

Figure 5. The pitch contour of Adnan Syed's first lines from the first *Serial* episode, "The Alibi." Serial's first episode, with pitch represented logarithmically on the *y*-axis scale, volume represented by the waveform at the bottom along the *x*-axis. When Syed says the word "exist," we see a vocal crack in the form of a drift upward. Decoded using Gentle and Drift. Image by the author.

ization process and therefore represents a space of risk and exposure.[45] In this context, a feeling of exposure is also present, and I would contend that the pitch shift also signifies in at least two more ways. First, the vocal crack makes Syed more "bodily" present, giving his voice what Roland Barthes called "grain," in the sense that we are unlikely to pass directly through the sound of his words toward their meaning without first having a certain amount of perception of friction taking place between the language and his vocal organs, a sense of impeding bodily presence.[46] Sound theorist Michel Chion links such a sense of presence—his term is "materializing sound indices"—with unevenness, mistake, and irregularity; it is significant that this roughness takes place in vocalizations by a racialized body, setting Syed apart from voices with the privilege of "identity-neutral" White radio voices that guide the narrative.[47] This, along with the poor sound quality of his prison phone line and the presence of inmates behind him, makes Syed's voice seem the most located-by-materiality of any voice in the podcast. It's not just that his larynx affects the sound but that the laryngeal slip seems to turn on a light, illuminating a network of interactions between the social protocols and media systems of the podcasting and penal worlds that dapple the edges of sound.

Second, since this sort of sonic mark is often associated with pubescent young people who present as boys, it also makes Syed come across as adolescent and developmentally ambiguous, even though Koenig emphasizes that at the time of her interviews he is a large, bearded man. This is important to the overall rhetoric of the show because it makes him seem temporally set off from everyone else we meet. All of the former teenagers whom we hear Koenig interviewing over the episodes of the season have grown older and weary. Syed's voice, by contrast, is preserved in amber as if still in that moment in 1999 when the murder took place. There is a paradox to the vocal crack, then, making the voice feel more material in present space but less material in present time. There is a common tonal denominator, however, as both valences of the vocal pitch shift make him seem vulnerable; I would wager this is one of the subconscious elements that made many listeners sympathetic to his account of events.

Where can the analysis go from here? The temptation to read pitch shifts as "tells" of hidden guilt is strong, and once you begin tracking it, odd patterns seem meaningful. In the sixth episode, for instance, Syed's voice cracks in exactly the same places when he repeats himself, discussing what he thought about when the police first contacted him to say Lee was missing: "The only thing I really associated with that **call** was that man, uh, you know **Hae**'s gonna be in a lot of trouble when she gets home [. . .] to me, all this **call** was, **Hae**'s going to get in a lot of trouble, you know, her mother is going to be pissed when she comes home, right?" But what does this doubling really "prove"? In this case, does the precision of a repeated pattern of shift mean that he's hiding something or that he is more forthcoming with his emotions than usual? Could it be that he's dehydrated (that can lead to voice cracks, too), or is it especially loud in the phone room at his western Maryland prison, and the strain manifests as a vocal crack?

One problem with this line of thinking is that it imputes the pitch shift entirely to Syed's psyche, unconscious or otherwise, but that is a mischaracterization of the interpretive situation that we face. Because this is a produced piece of narrative audio, one with many audible and inaudible edits of Syed's vocal appearances on tape, and indeed with larger questions about where to include his voice at all (in three whole episodes of the season he is not vocally present), we have no way of knowing how frequently or under what circumstances cracks really occur in his unedited voice or of separating the pitch shifts he made from the shifts the editors chose for the episode, perhaps for other reasons. Without access to the "raw" tape, the task of analyzing any particular element of his voice is set inside the task of analyzing how this voice is edited, refined, and portioned for our ears. And this does not even address how these cracks are tacitly framed by the ethical choice to interview him on tape in the first place. Had there been no recorded calls between Sarah and Adnan, no cracks would have occurred, because the words would never have been uttered. Syed chooses his words presumably with volition, and if these cracks were "tells," they could penetrate that performance with the proverbial Freudian slip, but the podcast editors also choose his words with their own set of needs and ideas, conscious or otherwise, so the "Freudian slip" could be betraying something about their feeling as well. The editors are speaking just as clearly as Syed through these cracks, so in a sense it is the podcast as a whole whose "throat is catching."

At least it is possible to come to grips with the pitch shift numerically. By my count, Syed makes eighty-five total vocal cracks across the nine episodes in which he appears vocally in the twelve-part season (see fig. 6). Remarkably, these eighty-five are distributed roughly evenly between three broad episode categories: those that end with things looking good for Syed's innocence (episodes 1, 9, and 11, a total of twenty-five vocal

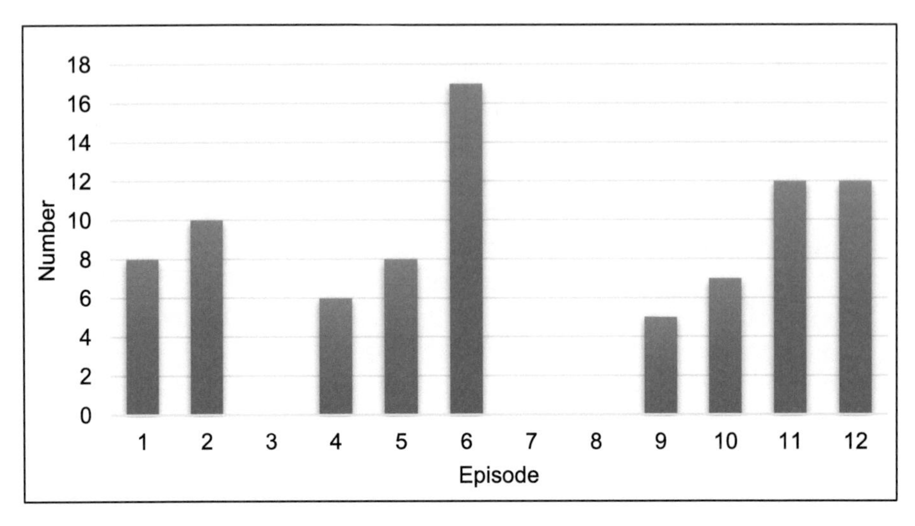

Figure 6. A graph of Adnan Syed's vocal cracks, by episode

cracks); those that end with things looking bad for him (episodes 2, 4, and 5, a total of twenty-four vocal cracks); and those that end in a mixed or neutral state (episodes 6, 10, and 12, the lion's share, a total of thirty-six cracks). This distribution neutralizes this overdetermined sonic element, to constrain its capacity to "mean something" coherent when it comes to a hermeneutics aimed at discovering tacit confession. But, using this map, we can hear that in situ many of the appearances of the crack lend themselves to specific interpretation. For instance, in episode 9, which concerns Syed's arrest and life in prison, a crack takes place whenever he is quoting his past self or describing thoughts that he was not able to express at trial ("You want to say, '**Hold on** but that's not true, that's not **the reason** why I got a phone'"), suggesting a search for rhetorical force that eludes him. There are also a few key words that have cracks suggesting genuine surprise when Syed first heard about them, such as "**Diary?**" referring to Hae's diary, which was brought forward at trial, and "**Jay?**" the first time the name of his main accuser comes up in the story. In episode 10, vocal cracks underscore his reverence for his failed attorney, Christina Gutierrez, who had since passed away ("I **loved** her") and also when Syed is taking Koenig to school about the reality of convictions in Baltimore County—"No one really **beats** cases, and when it comes to first degree murder cases it's almost **impossible**." In episode 12 there is a highly unusual vocal crack, one that sounds different from all the others, as Syed speaks of filing a motion to have forgotten swabs from Lee's body tested for DNA. "You guys had it sitting for sixteen years and you never **tested** it?" he says, choking up.

An even more interesting map comes into view if we focus on Syed's longer passages of dialogue. In episode 5, "Route Talk," the cracks are dense in a section in which he puts forward his theory that he could not have left school early enough to fit the prosecutor's timeline of the murder, because traffic would snarl at that time of day: "You **can't** just go to your car and **leave**. It's gonna **take** a few minutes. [. . .] From what **I** can **remember** those **buses** didn't **clear** in five minutes." In this case, it seems his voice cracks out of hope; this issue is something he had long chewed over and finally had the chance to prove through Koenig's re-creating the route. But many of Syed's dialogue passages also show the opposite. In his episode 9 story about how he was arrested in the first place, there is just one vocal crack in a 619-word description of his memories from the moment he was captured to the moment he asks for an attorney. It is an emotional passage, too, with Syed speaking about feeling shame and concern about his father. The same is true in the episode 2 passage with Syed's in-depth description of the day of the murder in 1999, which contains almost no cracks at all. Contrast these straightforward vocal performances with this passage in episode 11, which partially deals with a story about Adnan stealing money from the collections at his mosque:

> I mean, and it's a very uncomfortable thing for me to **talk** about, you know what I'm saying? [. . .] It's a very shameful thing that I **did**. I've never denied it. I don't see, I don't understand. I just think it's really **unfair** to me. [. . .] So it's **put** me in a **predicament** like, it's like you're **basically** publicly shaming me for something that I've never denied that I did, anyway. And it has nothing to do with the **case**. But you won't do it to other people though; it's like why do I have to keep getting called out on my stuff and it's got nothing to do with the **case**, but you don't do it to **nobody** else. You don't do it to **nobody** else.

In that same episode, Syed's voice cracks often when he says that he takes cold comfort in moral support: "For you to say that I'm a **great** person, to say that I'm a **nice** person. I only talked to you on the **phone** a few times."

Locally, then, the crack can have any number of indications—it's not a specific sign but a kind of metalanguage that conveys different affective tones, semantic references, and states of mind that come to us only dimly. Still there is a pattern that lends some insight on the series as a whole. Whenever Syed speaks of the elements of his case, he is likely reciting something that he has said many times before, and so it comes across as memorized or scripted if only by virtue of being repeated for decades. Recited as if by rote, these passages exhibit virtually no cracks. By contrast, nearly all of his most cracked sentences have to do with matters that bother his thought or conscience *in the present*, like the mosque theft that has come to light or issues and developments that pertain to the state of the case today. In the final episode, a third topic tends to involve a lot of pitch shifting—that of the podcast itself. In the opening of the last episode he speaks directly to Koenig about this:

> **You** don't really have, if you don't mind me **asking** you don't really have no ending? [. . .] I don't think you'll **ever** have 100 percent or any type of certainty about it. The **only** person in the **whole** world who can have that is me.
>
> I think you shouldn't really **take** a side. I mean, it's obviously not my decision, it's yours, but if I was to be **you**, just go down the **middle**. Obviously you know how to narrate it, but I checked these things out and these are the things that look bad against him, these are the things that the state doesn't really have an **answer** for. I think in a way you could even go point for point and in a sense you leave it up to the **audience** to determine.

In sum, then, it is clear that the "Adnan Syed" that the podcast shows us sonically (whether this is the "real" Syed is a separate matter, one for which the podcast is poor evidence) is at his most materialized, his most trapped and childlike, whenever he is responding to questions about how other people view him in his life and especially, by the time the season ends, how other people view him in the podcast. If the crack is a sign of monomaniacal concern, then it is about his feelings concerning his reputation and newfound celebrity. Inasmuch as the vocal crack "belongs" to Syed, then, the embodied nervousness it conveys attaches more to his concern about the way people out in the world are thinking about him than to his statements about the crime.

Shift these same attributes from "Adnan Syed" and locate them instead in *Serial* as a vocalizing entity, and the story is a little different. Inasmuch as the vocal crack "belongs" to *Serial* itself, it lets slip nervous energy about how the show's astoundingly broad and garrulous public of obsessives receive the piece. The crack is a sign of the show's sonic text interiorizing its own context as a podcast hit. In this way, the vocal crack captures in real time precisely *Serial*'s anxious obsession with its own unexpected success. Take the obsessive's path into the vocal crack, its materiality and odd childishness, and it leads to a network of micro-aesthetic moments that together generate a mood of bashful self-consciousness, even awkwardness and fear for its own dignity. As time goes by from episode to episode, the map of pitch shifting reveals the podcast to be an expressive project "doing" obsessive thought as it becomes aware that it has transformed into an object of obsessive thought, and in this way it brings the temporality of its release schedule into the time-space of its structure.

Obsession and Time

At the outset of this chapter, I proposed that to historically situate the rise of obsession in podcasting in the 2010s we can look back to the forgotten historical functions of monomania in nineteenth-century literature, where it represented a response to changed conditions that surrounded the advent of modernity with its fundamental shifts of everyday experience. The fascination with idées fixes today is similar in some ways, in that it represents a reaction to the omnipresence of feed-based media, be it social media or streaming services of various kinds in everyday modes of media consumption. Monomania both resembles the narcissistic hyperspecialization that online media automate and at the same time represents a mode of comportment that feels like it wrests control from automatic forces back into one's own internal mental churn. The obsession vogue also fits in with the notion of a genre and industry in a period of transition. Indeed, the concept of monomania may have done for the audio industry the same thing it did for early psychiatrists in the nineteenth century. At that time, as Jan Goldstein has argued, monomania as an idea helped the new profession of psychiatry negotiate "boundary disputes" between politics and medicine, between psychology and law, establishing its role in the landscape of modern professions.[48] For podcasters, the era of monomania helped creators reach fans, try out funding models, and rethink how audio content could be structured during a time of negotiation between legacy media companies, publishers, tech upstarts, and online platforms. Like monomania, obsession is a form that also began to fall from favor almost as soon as it served its function; even *Serial* retreated from obsession as a guiding idea in its later seasons.

The obsession vogue would have a legacy, though, by touching off a still ongoing question about time and structure. In the late 2010s, many podcasters liked to say that the new medium afforded liberation from the strictures of a broadcast schedule, with its insistence on specific increments of storytelling duration and frequency. Decades ago, stories were shorter, they had to fit into predefined time slots, and their production time's workflow had to obey the time schedule of a live feed from a radio station, according to a long-standing and tyrannical apportionment system of time, one that was never designed to serve storytellers or their public but instead to accommodate federal regulations and the requirements of commercialism. Podcasting, by contrast, seemed to have no externally imposed map delimiting length, segmentation, release schedule, or anything else when it came to the overall rhythms of production and expression.

But that didn't mean many podcasts were truly "free-form." On the contrary, it was in many cases obsession itself that took on these very mapping functions. An episode, a season, or a whole podcast would last exactly as long (funds and sponsorship allowing) as the obsession of its creators in the subject continued, at least according to its narrative. *Serial*'s first season runs about half the duration of Wagner's entire Ring Cycle, but at the conclusion only the mental churn of its host is really at an end; Syed's efforts to vindicate himself continued for years until his release in 2022. Indeed, you could think of many crime podcasts as a monomaniac's course of self-treatment rather than as a justice narrative in the mold of the crime procedural. An episode or season is structured to "solve" an obsession with a crime, or to let an obsessed reporter reach a state of equilibrium with their thoughts, rather than solving the actual crime, which in this case may have been solved all along and, if not, is still only a red herring for what amounts to a psychological journey.

Obsession mapping is thematically prevalent *in* podcasts of the 2010s because podcasts of the 2010s *were* obsession maps, structurally speaking. Shows used a distribution of affect to incrementalize narrative temporal framing, transmogrifying feeling into morsels of time. The conclusion of a season therefore feels so decisive. No matter how much a host professes their life-long obsession with their subject, this seemingly all-consuming feeling lasts no longer than the release structure permits, rarely spilling out of the structure devised to contain it. Hence, nearly all season-based shows follow just one obsession per season, and all anthology shows map just one obsession per episode, an oddly tidy way of compartmentalizing an emotional state that by definition defies compartmentalization. Small wonder that, when it comes to narrative podcasts, the first season is often the most vivid. Great radio shows tend to improve over years and even decades, like deep friendships, while in narrative-based podcasts later seasons seldom compare to the initial one, like first love. The end of an obsession is also the end of the space-time it had to invent in order to achieve expression; scheduling a second season on a new topic is like saying to the listener that actually the podcast producer has a whole other deep passion that is just as meaningful as the first, a claim that is at best undermining of the whole rhetoric of the first season and at worst simply ridiculous. When the thrill is gone, it might be more dignified to leave it alone.

If "obsession time" would not work as a replacement for "broadcast time," then what would? *Serial*'s subsequent seasons show self-consciousness about this, and both tried to find an alternative kind of structure that would help with the problem. The 2015 season of the program represents the first attempt to find a way out, focusing on the case of Bowe Bergdahl, an American soldier who left his post in Afghanistan, supposedly to trigger a search to bring to light uncomfortable truths about the leadership of his unit, but was captured and held prisoner by the infamous Taliban Haqqani network from 2009 to 2014. Koenig never speaks directly with Bergdahl on tape, and the show relies instead primarily on interviews with Bergdahl made by filmmaker Mark Boal. In place of obsession, the show anchors itself in its impressive resources. "The Golden Chicken," the second episode, draws on at least twelve interviewees from Bergdahl's battalion, twenty-five hours of taped conversations between Boal and Bergdahl, an interview with Taliban leader Mujahid Rahman, reporting by Afghan Sami Yousafzai, who made contact with a fighter named Hilal, as well as documents from WikiLeaks, all to reconstruct what happened to Bowe in the first weeks of captivity. Koenig clearly has sympathy for her subject. In the fourth episode, detailing his captivity, there is horrific awe to her description of a captor cutting Bergdahl's chest slowly with a razor blade hundreds of times. "Any one piece of this story can keep a person's mind churning," Koenig reflects. Yet the podcast is always moving forward, not resting in the churn in the way the previous season almost always did.

As a result, it is often hard to define what the season's central question is, as episodes cast about for some common denominator with the potential to connect the levels of the world it explores, from the life of a grunt stirring detritus in a toxic burn pit at an isolated outpost to horrifying battlefield experiences and Kafkaesque geopolitical complexities. Is the show about why Bowe went off base in the first place, whether anyone died looking for him, why it's all become so political? Unlike in the first season, the producers really do seem "fascinated" rather than "obsessed," and it is not clear that they find the quiddity of their own interest on which to rest the material. Unable to settle on a conundrum, and without the obsessiveness needed to map it into a structure, the season concludes with a portrait of the hauntedness of soldiers of the Afghanistan war, Koenig remarking, "These guys who deployed to war zones are so young, and then they're traumatized. All of them, pretty much. And it sucks." *Serial*'s second season ends up as something awfully familiar to listeners of *This American Life*, with its frequent interest in accidents, misunderstandings, ironies, and pyrrhic victories that merit rumination.

The third season takes a different route, reintroducing Koenig into the space of sound capture in a way we didn't hear in 2015 but shifting methodology again to an extended version of a different *This American Life* format, one that resembles ethnography. Some precedents include *TAL*'s "129 Cars" about life behind the scenes at a car dealership, "24 Hours at the Golden Apple" about a Chicago diner, and the "Harper High" series, all of which use a pinpoint location as a structuring conceit, adopting open-ended participant observation as an approach. In this case, it's the criminal courthouse in Cleveland, Ohio, where along with Emmanuel Dzotsi, Koenig rides the elevator up and down the floors to experience shoulder-to-shoulder closeness, musing how a government building can seem to both dispel and highlight the inequities in American life, particularly when it comes to a criminal system, where 96 percent of all resolutions come from plea bargaining, often to the detriment of people of color.

The soundtrack reflects this attempt at a diversity of perspective, drawing in classical, techno, and light jazz. The season is also the most sonically rich of the entire series, visiting more depicted locales and using more on-location recording. We hear about a young woman who accidentally hits a cop in a rowdy bar fight; listen to convicts flattering infamous Judge Daniel Gaul's ego to get a better sentence; hear activist Clevelanders in the wake of twelve-year-old Tamir Rice's death at the hands of police; join Koenig's debate with defiant police union head Steve Loomis; eavesdrop on the discussions in the office of Cleveland prosecutor Brian Radigan as his team decides on how to charge offenses, possibly all staged for Koenig's benefit; and eventually come to the story of Davon Holmes, falsely maligned and accused of the murder of infant Aavielle Wakefield. The Holmes accusation spans several episodes and nearly becomes the "main obsession" of season 3. Then the piece pivots, finding several other characters caught in a racist, sloppy, and ineffective justice system, including Jesse Nickerson, the victim of a savage police beating, and Joshua, a juvenile informant blackmailed by police to inform on gang members, who attack him regularly in a juvenile detention that is supposedly a model for the country.

Perhaps the most radiophonic moment in the show comes at the end of this season, with its extended multi-scene appraisal of the Cleveland justice center in a series of brief simultaneous scenes with a flat, vignette-like structure exploring every nook and cranny in the center from the arraignment room to the restrooms. This style strongly resembles the work of early feature makers such as D. G. Bridson and Norman Corwin.[49] Just as Corwin's works would whoosh his listeners around towns, cities, and countries to give social and audible form to specific occasions, segueing kaleidosonically from one fixed aperture to another under the guidance of a narrating voice as if simultaneously, *Serial*'s third season spins from courtroom to courtroom, catching the texture of life, from the terms of art used on the twenty-first floor to legitimately accept a plea bargain ("knowingly, intelligently, and voluntarily") to a scene of the lone Black woman on an otherwise all-White jury, expressing her unwillingness to participate due to her personal disappointment in the inequities of the system, down to the ground floor where Evin King is released after being exonerated, having spent twenty-two years in prison for a murder he didn't commit, then over to a march in honor of street cops who died on the job, then back into the center to hear a judge on the eighteenth floor lunge at a defendant. *Serial*'s true subject is how the Cleveland justice center embodies the space-time of police and judicial violence toward people of color. The metaphor of the building somehow manages to both contain that idea and leave it strangely unsaid.

Taken as a whole, across all three of its seasons, *Serial* is at its core an experiment in seriality that bridges the models of both public radio and podcast presentations, and it is therefore no surprise to see it wrestling with issues of temporality and space thanks to this double presence, one across a defined grid of time enforced by public broadcasting standards and strictures and another across an undefined time-space where the piece is at once without dimension and also permanently new in the archive. That later seasons seem to wander signals how complex a task it is to take the idea of the hourlong feature in the style of *This American Life* and re-temporalize it for a modern serial listening experience across many hours with a long release structure, particularly doing so as the reporting is still going on.

In the theory of narrative time, the oldest trick in the book is to develop a reading that maps the time of the *fabula* (the narrated events in chronological order) and the *sjuzet* (events in the order presented to the reader). The *sjuzet* has controlling power, as the *fabula* only becomes available to the listener through it. Because a *fabula* and *sjuzet* can never have an identical temporality, disjuncture between them produces narrative rhythms—you can, for instance, describe a "summary" of events as a passage in which the time of the *fabula* is longer than the time of the *sjuzet*; summaries often link together "scenes," which are spans of text in which the time of the *fabula* events and the time of their storytelling seem to equal one another. Drawing on the work of Gérard Genette, theorist Mieke Bal suggests five basic tempi in narrative writing: ellipsis, scene, summary, slowdown, and pause.[50] All this makes sense for radio as well, and a show like *This American Life* lends itself to meritorious assessments precisely because it can be rewardingly unpacked in these terms. *TAL* is particularly adept at slowdowns, taking time out of narration to signpost the importance of something about to happen.

But when it comes to narrative podcasting, there is a third temporal register to account for, since in addition to the time of the *fabula* and the time of the *sjuzet*, there is also the time of reporting, which is also taking place as the narration occurs, in between segments of the released narrative episodes. Thus, to put it in terms more common in media studies, the ellipses, scenes, summaries, slowdowns, and pauses that represent temporal reconfigurations of the *fabula* by the *sjuzet* are themselves governed by ellipses, scenes, summaries, slowdowns, and pauses dictated by the time of reporting and how it influences (and is influenced by) the structure of the *sjuzet*. In effect, where the analysis of narratives traditionally has to build their understanding of temporality by looking to discrepancies between the time of narration and the time it depicts, the analysis of podcasts like *Serial* also needs to account for a time of composition that affects both and is itself, like the *fabula*, only hinted at by the *sjuzet*, as we only even know about the time of reporting because it is relayed to us. This triple layering of time comes to the fore most strongly whenever podcasts—*Serial* is one of them—feel the need to add episodes months or years after the initial run, revealing the way the text represents the intersection of the triple timelines of events, reporting, and narrating.

It is clear why temporality was such a complicated aesthetic problem for podcasters who told stories as they reported them and faced no externally imposed rules about the when of release or the length of expressive depiction. This was both a blessing and a curse. At first, the solution became obsession, which decided not only the raison d'ètre of the show but also where the "origin" point is for its narrator. It also helped shape when episodes should end or begin, how many there should be, and when a season felt "over." Ever since that first season, however, the *Serial* team retreated from this approach. And while subsequent related shows such as *S-Town* and *The Trojan Horse Affair* would personalize stories in other ways, the obsession with obsession was neither sustainable nor replaceable. If the Syed season represented an attempt to replace the dictates of broadcast time with obsession, then the Bergdahl season represents an attempt to replace it with story scale and the Cleveland season an attempt to replace it with place. The fit is never quite right, the map always exceeding the territory or the territory the map.

As time went on, of course, the podcast networks themselves began to do what broadcast networks and stations once did, streamlining podcast durations and frequency of release. Using podcast metadata, the University of Wisconsin's PodcastRE project has uncovered clear evidence of house styles in this respect—according to their research, Earwolf podcasts had wildly different durations, while Gimlet shows had relatively uniform ones.[51] The very fact of this heterogeneity shows the complexity of the problem. Never does podcasting seem more in its infancy than in the way it grabs for readymade temporal structures like "seasons" and "pilots," unable to devise temporal units that adroitly reflect the form. Never does it seem dreamier, as it has in more recent years, than when it slows down unexpectedly, disengages, and modulates in unexpected ways. *Slow Radio*, a BBC 3 podcast, has since 2017 been distributing deliberately meandering place-based audio, including river journeys, sound walks, monk meditations, and work by master nature recordists such as Chris Watson. *New Yorker* critic Sarah Larson described it as an "ideal blend of companionship and earthly randomness."[52] *Rumble Strip*, a Vermont-based podcast, has been another interesting one to follow in this area, a podcast that conveys a sense of patience in its stories of rural waitresses and workers in barns, one that also consistently focuses on oral histories of the disabled. In 2020's "I Am In Here" host Erica Heilman interviews Mark Udder, a man with a form of autism that prevents him from speaking and who has only recently been able to use a speech-to-text software program to verbally articulate his thoughts. The podcast makes a point of letting the time of Udder's typing elapse on tape rather than truncating its duration to accommodate the listener's impatient temporality, as he reflects both on what the software has given him and on what it has taken away.

Shows like *Rumble Strip* represent an early form of what became the slow audio movement during the pandemic years. A rejection of the need to explain and define "slow audio" was a topic of several podcasts and radio shows: one of the first *Invisibilia* episodes when it relaunched in 2021 with hosts Yowei Shaw and Kia Miakka Natisse was titled "The Great Narrative Escape," prompted by the project of what it means to make a "boring" show. Producer Eleanor McDowall of Falling Tree productions made or curated several works that explored the theme of slow time, from her podcast *Field Recordings*, which featured audio makers recording silently in fields, to pieces produced for Josie Long's *Short Cuts* and *Lights Out* on the BBC, venues that explored slow audio consistently in several works that ask questions about the very nature of making audio. In 2022, artist and producer Ariana Martinez synthesized many of these practices in "A Time-Based Medium," an essay on the Transom website that highlighted how McDowall and several others were using long wordless field recordings, slow fades, and durations that play with the form of the "audio postcard" and last the duration of a "long walk."[53] Martinez put their finger on a pressing need for audio to work on fresh concepts of internal pace, to build slowness into how interviews, workflows, and edits functioned, and thereby to truly reflect on what it means to be a time-based medium rather than to reflexively grab, push, and pull the audience through a story, out of old habits. Audio should be more like textile work, more like carving, they argued, an approach that resists both the impatience of traditional radio, with its insistence on immediate hooks, and the obsessiveness of long-form podcasting, with its pursuit of monomaniacal engrossment.

You can imagine the rise and fall of "obsession time" as an aesthetic of temporality that thrived in between the conventional temporality associated with linear radio (where, whether live or taped, everything has to happen at a rhythm that suits a fixed and unalterable broadcast schedule), and a much more experimental and deeper practice of questioning time, like the kind currently being explored in the slow audio movement. If that is right, then obsession temporality appears less like a stable model and more like a transitory disruption whose brief vogue illustrated that something had changed about how audio was "thought." There was, in these years, a break with normative approaches to audio temporality that affected radio, podcasting, and the forms that stood between them. Whatever linear radio's problems were, over its century of regulation, deregulation, and commodification, the medium had learned what to do with time as a constraint to its inherent plasticity, how to portion and subportion it out, institute rhythms that extract value, expectations, and clarity, sometimes so much so that we lose a sense of the inherent strangeness of the space-time of spectrum-based media to begin with. Obsessive time was a step toward rediscovering that strangeness, but it was a stumbling step. The problem of understanding and deploying podcast time, at least for those who wrestled with this issue in the late 2010s, was never fully worked out, and by the dawn of the third decade of the millennium, the problem of podcast time remained—frustratingly for some, tantalizingly for others—unresolved.

Audio Works Discussed in Order of Appearance in the Text

Serial (Sarah Koenig, Julie Snyder, et al., WBEZ Chicago, Serial Productions, October 3, 2014–present, https://serialpodcast.org/).

Slate's Serial Spoiler Specials (David Haglund, Julia Turner, Katy Waldman, Slate Podcasts, October 28, 2014–September 24, 2022, https://podcasts.apple.com/us/podcast/slates-serial-spoiler-specials/id935063801).

This American Life (Ira Glass, WBEZ Chicago, November 17, 1995–present, https://www.thisamericanlife.org/archive).

Reply All, "The Case of the Missing Hit" (Alex Goldman, PJ Vogt, Gimlet Media, March 5, 2020, https://gimletmedia.com/shows/reply-all/o2h8bx).

In The Dark, Season 2: *Curtis Flowers* (Madeleine Baran, APM Reports, May 1, 2018–June 11, 2020, https://features.apmreports.org/in-the-dark/season-two/).

Reveal (Al Letson, Center for Investigative Reporting, PRX, September 28, 2013–present, https://radiopublic.com/Reveal).

Between the Ears, "Slow Movement: Everything, Nothing, Harvey Keitel" (Pejk Malinovski, Falling Tree for BBC Radio 3, October 12, 2013, https://podcasts.apple.com/us/podcast/slow-movement-everything-nothing-harvey-keitel/id721125509?i=1000345777078).

The Documentary Podcast, "Dreaming Dickens" (Cathy FitzGerald, BBC World Service, February 3, 2012, https://www.bbc.co.uk/programmes/p02rrz63).

In One Ear and Out the Other (Tim Hinman, ABC Radio National, November 8, 2014, https://www.abc.net.au/radionational/programs/archived/radiotonic/in-one-ear-out-the-other/5861876).

The Heart (Kaitlin Prest, Mermaid Palace, Radiotopia, February 5, 2013–February 23, 2022, https://www.theheartradio.org/all-episodes).

Rabbits (Terry Miles, Public Radio Alliance, February 28, 2017–March 15, 2022, https://www.rabbitspodcast.com/episodes).

Love + Radio (Nick van der Kolk, October 18, 2005–present, https://www.wbez.org/shows/loveradio/700796df-7723-4ace-9c80-3121ca615605).

The Generation Why Podcast (Justin Evans, Aaron Habel, Wondery, 2012–present, https://wondery.com/shows/generation-why).

Last Podcast on the Left (Ben Kissel, Marcus Parks, Henry Zebrowski, Last Podcast Network, March 29, 2011–present, https://www.lastpodcastontheleft.com/).

Sword and Scale (Mike Boudet, Incongruity, January 1, 2014–present, https://www.swordandscale.com/).

Criminal (Phoebe Judge, Radiotopia, 2014–present, https://thisiscriminal.com/).
Slow Burn (Leon Neyfakh, Andrew Parsons, Slate Podcasts, November 28, 2017–present, https://slate.com/podcasts/slow-burn/s7/roe-v-wade).
Crimetown (Marc Smerling, Zac Stuart-Pontier, Gimlet Media, November 2016–March 11, 2019, https://gimletmedia.com/shows/crimetown/episodes#show-tab-picker).
Ponzi Supernova (Steve Fishman, Audible, May 5, 2017–June 9, 2017, https://ponzi.libsyn.com/).
Uncover: Escaping NXIVM (Josh Bloch, CBC Podcasts, September 3, 2018, https://www.cbc.ca/listen/cbc-podcasts/187-uncover).
Oversight: Jonestown (Sheila MacVicar, CQ Roll Call, November 17, 2019–December 22, 2019, https://luminarypodcasts.com/listen/cq-roll-call/oversight-jonestown/f8ac715d-5c92-4ece-8709-a846946f498c?country=US).
Guru (Ginny Brown, Matt Stroud, Laura Tucker, Wondery, June 15, 2020–July 27, 2020, https://wondery.com/shows/guru/).
Mogul: The Life and Death of Chris Lighty (Reggie Ossé, Loud Speakers Network, Gimlet Media, June 11, 2017–June 15, 2017, https://open.spotify.com/show/4okzhZKxw5lSXcADxVxNRj).
Chapo: Kingpin on Trial (Keegan Hamilton, VICE News, November 1, 2018–December 17, 2018, https://open.spotify.com/show/3iZGZfoQX9kfzdZtAYi2s2).
100:1 The Crack Legacy (Christopher Johnson, Audible Originals, October 26, 2017, https://www.amazon.com/100–1-The-Crack-Legacy/dp/B08DDFKNWY).
The Uncertain Hour (Krissy Clark, Marketplace, April 28, 2016–March 24, 2021, https://www.marketplace.org/shows/the-uncertain-hour/).
The Home Babies (Becky Milligan, BBC Radio 4, June 15, 2018–July 27, 2018, https://www.bbc.co.uk/programmes/p06b5fzl/episodes/downloads).
Making of a Massacre (Ginger Thompson, ProPublica, Audible Originals, May 3, 2018, https://www.amazon.com/The-Making-of-a-Massacre/dp/B08DDF1MYM).
Ear Hustle (Nigel Poor, Earlonne Woods, Rahsaan "New York" Thomas, Radiotopia, June 14, 2017–present, https://www.earhustlesq.com/listen).
Bird's Eye View (Johanna Bell, StoryProjects, March 7, 2020–April 9, 2020, https://www.birdseyeviewpodcast.net/podcast).
Prison Bag (Josie Bevan, Rebecca Lloyd-Evans, Alan Hall, Falling Tree, April 24, 2020, https://podcasts.apple.com/us/podcast/prison-bag/id1510319335).
Under the Other Pain (Charlotte Rouault, RTS Le Labo, Faïdos sonore, September 16, 2020, https://ifc2.wordpress.com/under-the-other-pain/).
70 Million (Mitzi Miller, LWC Studios, August 27, 2018–present, https://70milliopod.com/season-5).
Hollywood & Crime (Stephen Lang, Tracy Pattin, Wondery, January 6, 2017–November 3, 2017, https://wondery.com/shows/hollywood-crime/).
Out of the Blue (Michael Braithwaite, Blue Door, October 7, 2019–January 28, 2021, https://outoftheblue.libsyn.com/).
Inside Psycho (Mark Ramsey, Wondery, March 23, 2017–July 10, 2017, https://wondery.com/shows/inside-psycho/).

Stranglers (Portland Helmich, Earwolf, Northern Light Productions, November 16, 2016–February 15, 2017, https://podcasts.apple.com/us/podcast/stranglers/id1174116487).

The Vanished (Marissa Jones, Wondery, December 31, 2015–present, https://wondery.com/shows/the-vanished-podcast/).

Dirty John (Christopher Goffard, *Los Angeles Times*, Wondery, October 2, 2017–November 21, 2018, https://podcasts.apple.com/us/podcast/dirty-john/id1272970334).

Dr. Death (Laura Beil, Wondery, September 4, 2018–September 21, 2021, https://wondery.com/shows/dr-death/).

The Shrink Next Door (Joe Nocera, Wondery, May 21, 2019–December 17, 2021, https://wondery.com/shows/shrink-next-door/).

The Joe Rogan Experience (Joe Rogan, Spotify, December 23, 2009–present, https://open.spotify.com/show/4rOoJ6Egrf8K2IrywzwOMk).

You Must Remember This (Karina Longworth, April 16, 2014–present, http://www.youmustrememberthispodcast.com/episodes).

Shattered (Jeremy Allen, WDIV Local 4, Graham Media Group, April 14, 2018–January 7, 2020, https://www.clickondetroit.com/topic/Hoffa/#episodes).

What Really Happened (Andrew Jenks, Andrew Jenks Entertainment, Inc., Seven Bucks Productions, October 25, 2017–May 4, 2020, https://open.spotify.com/show/02MBaVBDYsnK6GawvKZSfe).

Up and Vanished (Payne Lyndsey, Tenderfoot TV, August 7, 2016–May 26, 2022, https://podcasts.apple.com/us/podcast/up-and-vanished/id1140596919).

Monster: The Zodiac Killer (Matt Frederick, Payne Lyndsey, Tenderfoot TV, HowStuffWorks, iHeart Media, January 2, 2019–January 23, 2020, https://podcasts.apple.com/us/podcast/monster-the-zodiac-killer/id1477386458).

Atlanta Monster (Payne Lyndsey, Tenderfoot TV, HowStuffWorks, iHeart Media, December 31, 2017–August 1, 2019, https://podcasts.apple.com/us/podcast/atlanta-monster/id1324249769).

West Cork (Sam Bungey, Jennifer Forde, Audible Originals, February 8, 2018, https://feeds.acast.com/public/shows/620c6279-d89b-4719-947d-a5f4b47b44eb).

Thunder Bay (Ryan McMahon, Canadaland Investigates, October 22, 2018–December 15, 2020, https://www.canadaland.com/shows/thunder-bay/).

Breakdown (Bill Rankin, *Atlanta Journal-Constitution*, 2014–present, https://www.ajc.com/news/breakdown/).

Murderville (Liliana Segura, Jordan Smith, The Intercept, December 20, 2018–April 23, 2022, https://theintercept.com/podcasts/murderville/).

Disappeared (Sara Anzari, Investigation Discovery, August 5, 2021–October 27, 2022, https://podcasts.apple.com/us/podcast/disappeared/id1577613725).

Casefile True Crime (Mike Migas, Casefile Presents, January 9, 2016–present, https://casefilepodcast.com/episodes/).

Small Town Dicks (Yeardley Smith, Paperclip Ltd., August 2017–present, https://www.smalltowndicks.com/episodes/).

The Accused (Amber Hunt, Amanda Rossman, *Cincinnati Enquirer*, September 7, 2016–present, https://accusedpodcast.com/).

My Favorite Murder (Georgia Hardstark, Karen Kilgariff, Feral Audio, January 13, 2016–present, https://myfavoritemurder.com/episodes).

Wind of Change (Patrick Radden Keefe, Crooked Media, May 11, 2020–October 11, 2022, https://crooked.com/podcast-series/wind-of-change/#all-episodes).

Caliphate (Rukmini Callimachi, New York Times Company, April 19, 2018–December 18, 2020, https://caliphate.simplecast.com/episodes).

Missing Richard Simmons (Dan Taberski, Stitcher, First Look Media, Pineapple Street Media, February 15, 2017–November 1, 2018, https://www.topic.com/missing-richard-simmons).

Dolly Parton's America (Jad Abumrad, Shima Oliaee, OSM Audio, WNYC Studios, October 3, 2019–December 31, 2019, https://www.wnycstudios.org/podcasts/dolly-partons-america).

Making: Oprah (Jenn White, WBEZ, November 10, 2016–December 30, 2016, https://www.wbez.org/shows/making/71b8de57-b2be-4e03-8481-683258de3ec1/page/4).

Another Dead Man Walking (Ian Woods, Sky News, August 20, 2015–November 7, 2018, https://www.spreaker.com/show/another-dead-man-walking).

Revisionist History (Malcolm Gladwell, Pushkin Industries, June 16, 2016–present, https://www.pushkin.fm/podcasts/revisionist-history).

Bear Brook (Jason Moon, New Hampshire Public Radio, October 3, 2018–present, https://www.bearbrookpodcast.com/season-one).

Snap Judgment (Glynn Washington, Snap Judgment Studios, PRX, July 2010–present, https://snapjudgment.org/podcast-episodes/).

Heavyweight (Jonathan Goldstein, Gimlet Media, September 23, 2016–present, https://gimletmedia.com/shows/heavyweight#show-tab-picker).

Nancy (Tobin Low, Kathy Tu, WNYC Studios, April 9, 2017–June 29, 2020, https://www.wnycstudios.org/podcasts/nancy/episodes).

Mystery Show (Starlee Kine, Gimlet Media, May 21, 2015–July 31, 2015, https://gimletmedia.com/shows/mystery-show/episodes#show-tab-picker).

99% Invisible (Roman Mars, KALW Public Media, Radiotopia, September 23, 2010–present, https://99percentinvisible.org/archives/).

The Butterfly Effect (Jon Ronson, Audible Originals, October 24 2017, https://www.amazon.com/The-Butterfly-Effect/dp/B08DDCTB6V).

Heaven's Gate, "4: The Host" (Glynn Washington, Jenna Weiss-Berman, Pineapple Street Media, October 31, 2017, https://www.stitcher.com/show/heavens-gate/episode/4-the-host-200150609).

Mississippi Goddamn (Al Letson, Center for Investigative Reporting, PRX, October 16, 2021–December 4, 2021, https://revealnews.org/mississippi-goddam/).

Suave (Maggie Freleng, Futuro Studios, PRX, February 9, 2021–May 11, 2022, https://www.futuromediagroup.org/suave/).

S-Town (Brian Reed, Serial Productions, March 28, 2017, https://stownpodcast.org/).

The Trojan Horse Affair (Brian Reed, Hamza Syed, Serial Productions, New York Times Company, February 3, 2022, https://www.nytimes.com/interactive/2022/podcasts/trojan-horse-affair.html).

Invisibilia, “The Great Narrative Escape” (Kia Miakka Natisse, Yowei Shaw, NPR, May 20, 2021, https://www.npr.org/2021/05/19/998228413/the-great-narrative-escape).

Slow Radio, (Verity Sharp, BBC Radio 3, November 19, 2017-present, https://www.bbc.co.uk/programmes/p05k5bq0).

Rumble Strip, “I Am In Here” (Erica Heilman, Mark Uddar, Emily Anderson, October 29, 2020, https://rumblestripvermont.com/episodes/).

Field Recordings (Eleanor McDowall, Field Recordings, March 7, 2020–present, https://fieldrecordings.xyz/).

Short Cuts (Josie Long, BBC Radio 4, May 28, 2013–present, https://www.bbc.co.uk/programmes/b01mk3f8/episodes/player).

Lights Out (Various producers, Falling Tree, BBC Radio 4, December 12, 2022–present, https://www.bbc.co.uk/programmes/m000176d/episodes/player).

CHAPTER 2

Structures of Knowing

Something Out of Nothing

In the Ballyhoura Mountains overlooking Golden Vale pastures in Ireland's County Limerick, there is a village of some eight hundred residents called Kilfinane. Since 2014, the town has been irregularly hosting a gathering of radio makers called the HearSay Audio Arts Festival, a brainchild of radio presenter and local impresario Diarmuid McIntyre.[1] Once little more than a set of impromptu meetings in churches, pubs, living rooms, and storefronts, by the time I attended in the spring of 2019, HearSay was a four-day affair with more than one hundred participants from thirty-two countries and had become known as a kind of Radio Woodstock, with a flexible schedule, improvised lodging, logistical hiccups, scattered installations, and sound walks, all providing a cobbled together, muddy event in which audio dabblers could mingle with award-winning artists and industry figures in pubs to share tape about ideas and ideas about tape.

There were presenters from *The Heart* and *The Allusionist* podcasts, as well as debuts from *West Cork* and *Forest 404*. Producer Kara Oehler spoke about her work the previous year editing an all-sound edition of the *New York Times Magazine*'s travel issue that featured the sounds of rats chattering in Manhattan and of Hawaiian lava; leading radio memoirist Sophie Townsend described stories she would never make for air and why; and *The Stoop* host Leila Day told of producing "The Birth of Solomon" on the death of her sister's newborn, one of many children of color affected disproportionately by complications at birth. In a small room, groundbreaking radio translation group *Radio Atlas* did yeoman's work, screening subtitled English versions of hard-to-find international historical features, while the experimental podcast *Constellations* exhibited pieces that walked the line between radio and sound art. There were tiny stories installed in parked cars and prams, plant sonification shows, curated audio pieces paired with chocolate boxes, and a sonic pub quiz. There was even a dance.

HearSay 2019 took place at a key moment in pre-pandemic podcasting history. Narrative podcasts were no longer a novelty; while most attendees came from radio stations, national broadcasters, and production organizations in Europe and the Americas, among the remainder there were perhaps just as many attendees eyeing Audible contracts as there were penniless students. It had become clear, by then, that the pact between the public radio and tech sectors that launched the narrative podcast boom was driven more and more by the logic, needs, and models of the latter. It was also a moment of introspection for many in the audio world. The #MeToo movement had revealed that a number of hosts, producers, and other industry gatekeepers had histories of sexual impropriety and toxic behavior, leading to a string of firings, resignations, and revisions to codes of conduct, a process that continues today.[2] It was also six months after a pivotal speech at the Third Coast International Audio Festival in Chicago by producer Phoebe Wang. Accepting an award for her feature "God + the Gays" on *The Heart*, Wang took to the mic to upbraid the attending leaders of the industry for their hiring practices, which, for all their stated values promoting equity, failed even in these booming years to make meaningful effort to diversify staff and to reach the communities of color they serve.[3] Wang offered a new web resource she had helped put in place that lists public radio workers of color, so there could be no future "lack of applicants of color" excuse. These trends were each complex, and they had different origins and stakeholders, but they came together into a misty lack of clarity about what the commitments of this art form and industry ought to be. Audio's core philosophy, once firmly rooted in the pursuit of stories that foster empathy—perhaps a naive fantasy, perhaps an alibi for toxic workplace culture—seemed at that moment hard to describe, even among members of this rarified international group, and this uncertainty formed an emotional backdrop for events at HearSay and beyond.

Among the presentations I attended was a midday talk by Scott Carrier in a rain-soaked tent in a stone garden. A legendary award-winning veteran of *This American Life* and NPR, Carrier played a very beautiful piece from his podcast *Home of the Brave* that was based on a trip he took to Honduras, part of a series exploring what life is like in a country from which many were fleeing in what the media were calling "caravans."[4] The first episode of the series, "Honduras Part One," is a monologue in which Carrier relates an encounter with a woman and her children as they begin their dangerous journey north. He begs them to reconsider, having seen the risks, and reflects on his own privileged role in telling radio stories about vulnerable people. The second episode, "Tegucigalpa," offers a car tour of the titular city, exploring how corruption and territory shape life in the capital's troubled neighborhoods, known for their kidnapping and drug trades. The third episode, "Warriors Zulu Nation Honduras," the one that Carrier presented at HearSay, is about Chamelecón, a suburb of the city of San Pedro Sula, which was known at the time for its high murder rate and as one of the origin points of many migrants fleeing violence fomented by street gangs that were originally founded in Los Angeles. As Carrier's piece explains, although maligned by right-wing media, many Hondurans who emigrated did so under unimaginable duress, facing a choice to either join gangs or die.

In the piece, Carrier chooses not to focus on violence but instead to highlight an arts movement in the town, one whose flourishing in these circumstances might be surprising to his listeners, following local rapper Juan Carlos, a leader of the area chapter of the Bronx-founded international arts group Universal Zulu Nation. By virtue of being an artist, Juan Carlos was immune from intimidation and violence from MS-13 and Barrio 18, the two largest gangs in Chamelecón. As the two walk the streets, Carrier describes what he sees and ends up in a public park, where a community rap battle was planned that night. The imagery is memorable:

> [The park] had a tree that looked like a nuclear explosion hanging over a basketball court/soccer field. The sun was bright and the tree had lost its leaves, so the limbs cast shadows like blood vessels covering the court from goal to goal. The boom box was on the ground, near the trunk of the tree, just off the sideline at half quarter. It was a small speaker, a nine-inch tube, but it seemed like a magical tap pulling rhythm from the ground, from the roots of the tree.

In the remainder of the piece, we hear improvised performances by hip-hop artists and an excited crowd and learn more about the Warrior Zulu Nation in Honduras, where Juan Carlos struggled to recruit impoverished young people into dance and street art. Carrier concludes with his own reflection on this utopian scene as the "flip side" of the Trump administration's use of walls and detention centers on the border, policies all about "harboring fear and hiding from reality." Juan Carlos's practice was the opposite of all that, "about creating something outta nothing," a philosophy Carrier also claimed for his podcast.

Sentimentalizing a timeworn lesson about making works of beauty in circumstances of danger and deprivation, Carrier's parable of a gathering of hip-hop artists on a soccer field under a Honduran tree offered a strange parallel with a gathering of radio artists in the Ballyhoura hills. On the one hand, the story of Juan Carlos's group drawing subterranean rhythms from the earth provided the gathered crowd an invigorating parable about the vitality of art, while on the other hand it offered a reality check about the differing stakes between the Zulu Nation in one of the most dangerous places in the world and the podcast elite in one of the least. This idealistic episode also seems to know that its "lesson" is a wishful one, carrying with it an undercurrent of pessimism. Carrier's radio work has long been proximal to violent crime and trauma, especially involving migrant populations, but in this episode violence takes place largely off-tape, in the local and geopolitical background of the scene, as do the obstacles in making the piece itself. There are murky patches at the edges of the sharp picture; a mood of doom underneath is hard to miss. After all, while Zulu Nation rappers magically make something out of nothing, they do so on an empty field the color of blood, under a mushroom cloud.

Was there a similar feeling of impending doom in Kilfinane? In the "Woodstock of Radio," artistic idealism was alive that spring, but there were many ways in which the shadow of skepticism toward the kind of audio that Carrier's idealistic story and its habit of conveying knowledge executed so effectively had begun to creep in, both thanks to the encroachment of corporate interests and because of a growing suspicion about the aesthetic itself, developments that represent a Charybdis and Scylla in prestige audio even today. The Chamelecón episode was among the most vividly written and well-executed pieces I heard at HearSay that year, but it was also among the most old-fashioned.

An Epistemological Turn

This chapter explores how, in the late 2010s, many audio creators made work just like the *Home of the Brave* episode, exhibiting an urge to find something out about other places, time periods, people, belief systems, and societies. To do so, they adapted radiophonic narrative forms that had been popular for a generation and invented new ones to shape knowledge in story-based podcasts. If the last chapter turned on questions of how post-*Serial* "obsession" podcasters thought about feeling, this chapter is about how they felt about knowing, in particular how their work considers what we can and cannot comprehend, how a searching imagination faces impediments, how audio's mood sometimes contradicts its "lessons," and how loose ends could linger at the edges of stories that were told in a medium that was more open and shambolic than the tight, elegantly written radio features that admired creators like Carrier and his peers are known for. What follows is also about how narratives could be shaped, and modes of address altered, by cutting out or delaying knowledge in podcast settings. All narratives restrict knowledge, by necessity, and podcasting is no different, but they also can use restrictions to shape the process of apprehension. "Narrative always says less than it knows" theorist Gérard Genette has pointed out, "but it often makes known more than it says."[5] One of the key projects that podcasts of the late 2010s tended to undertake had to do with how to distinguish between what a podcast is trying to inform us about and what it is prompting us to interpret seemingly on our own. In this medium in this period, the varieties of restriction that proved to be aesthetically and ethically compelling, and those that were contested, tell us something about the evolving conscience of the field.

This chapter makes the case that, listening through the archive of podcasts of this period, both in the United States and abroad, you can hear a growing urgency of reflection around epistemology. While a cliché of the opening of many podcasts of the late 2010s was the ritualized confession of the narrator's "obsession" with the topic of the piece, as I argued in chapter 1, one of the clichés at the conclusion of podcasts was often about what is "known" by the end of the exercise, as the engine of obsession sputters out. The last episode of *Serial*'s first season, attempting to escape its own labyrinth of mental churn, is titled "What We Know." *Up and Vanished* uses that very title in its second season. So does the first episode of the fiction faux-true crime show *Limetown*. The last sentence in the last episode of the Norwegian historical mystery *Death in Ice Valley* is "someone out there must know something," a phrase that is almost verbatim the title of the CBC's entrance into the true crime podcast genre, *Someone Knows Something*. I explain the roots of this vogue for "knowing" in more detail below, but I should specify at the outset that it is not clear to me that what I designate as an epistemological turn in podcasts of the late 2010s was a direct choice on anyone's part. I am proposing the idea of a turn marked by epistemological characteristics as a *critical* approach, a heuristic for the development of strategies of close listening appropriate to many works of the period, rather than as strong evidence of a "movement" among practitioners like those at HearSay or anywhere else. I feel that some of the extravagances of late 2010s audio just make more sense if you approach these podcasts as stylized depictions of the acquisition of knowledge, as reflections upon the feeling of "coming to know" something, or as hard questions about what it means to *know* in an audio medium at all—this is how I am using "epistemology" here, a nontechnical usage that is nonetheless partially informed by the enduring fascination that formal epistemology has long had with the relationship between knowledge and constraint.[6]

Indeed, any listener to investigative podcasts from this period will be familiar with the way these podcasts cannot help but let us know many details about all the work that went in to finding out what the narrative will eventually array before us—the archives, the interviews, the far-flung travel, the years of labor away from friends and family, the dead ends. Many traditional radio shows required just as much production legwork, but they seldom made the dramatization of that legwork a part of the main text of the broadcast, committing the faux pas of drawing the listener's focus to the storyteller rather than to the main characters. In contrast, by the late 2010s few podcasts would try to distinguish themselves among their peers without making textually explicit a litany of the herculean trials of persistence through fatigue and tedium that their work required.

Meanwhile, these podcasts also aestheticized the process of "coming to know" for their audiences. When podcasts took on the venerable "human interest" story in these years, for instance, they tended to do so in ways that blend production aesthetics with oral history in a mutually integrating way, in works from *Nocturne*'s "A Hole in the Night," about a late-night trucker telling a story, related in between peels of steely music, about barely surviving a sudden sinkhole in the highway, to *Love + Radio*'s "Jack and Ellen," a story produced by Mooj Zadie about a young gay woman who poses as a young boy online to bait and blackmail pedophiles, her voice digitally disguised to be coded male for half the podcast in a fascinating piece of auditory trans-experimentation. In a review for the *RadioDoc Review*, podcaster Michelle Macklem pointed out that "Jack and Ellen's" musicality is not incidental but rather is at the center of how the producers communicate "editorial perspective."[7] Both "Hole" and "Jack and Ellen" use the mechanics of the emerging form—scattering reflective moments and musical stings, deploying artfully deceiving autotuned voices—to fill you less with empathy for their subjects and more with awe, astonishment, and sometimes active antipathy. In doing so, these stories stylize their own act of transfer of another person's experience, one in which an aesthetic framework is meant to contribute to the conveying of a fragment of life previously unknown to a listener. In this way, a design concept is interior to a process of "coming to know" what a story is about.

Some podcasts organize themselves to try to give their listeners a personal experience that had become dim in memory. Benjamen Walker's *Theory of Everything* did a marvelous episode on being a kid in 1984—"1984 (the Year Not the Book)"—humorously refracted through the "Where's the Beef?" ad campaign of that time. Other podcasts focus on national experiences that had become dim history. John Biewen and Chenjerai Kumanyika's works in this area are great examples, from reclaimed histories of Black life during the Civil War era, like the story of Shedrick Manego, hero of the Combahee Raid (*UnCivil*, "The Raid") in South Carolina, to the forgotten story of the largest mass execution in American history, the hanging of thirty-eight Dakota warriors in 1862 in Minnesota (*Scene on Radio*, "Little War on the Prairie"). A rash of podcasts attempted to make history richer or more fun (*Backstory*, *American History Tellers*), while others attempted to make it relevant to contemporary events (*Slow Burn*, *The Frost Tapes*, *Bag Man*, *Fiasco: Bush v Gore*), and several tried to reclaim it (*After Stonewall*, *Making Gay History*, *Shots in the Back: Exhuming the 1970 Augusta Riot*, *Louder Than a Riot*) from being lost. A small subset took an experimental approach. Nate DiMeo's *Memory Palace* has since 2008 been providing short, often ambiguous vignettes on historical topics, from the work of botanist Ynés Mexía, to the founding of the first cog railway on Mount Washington, to snowball fights on Boston Common. The Kitchen Sisters brought their prodigious archival audio research capabilities to the creation of *Fugitive Waves*, a podcast that in these years featured the archives of famed recordist Tony Schwartz, Sun Records producer Sam Phillips, and oral historian Studs Terkel, as well as a whole series called "The Keepers" about preservation passions, from the Hip Hop archive at Cornell University to the Cinémathèque Française. The Kitchen Sisters archive itself was recently acquired by the America Folklife Center at the Library of Congress.[8]

This wave of archontic podcast output about the past is an excellent example of Diana Taylor's concept of how tangible archives—documents, letters, records that resist change—can interact with repertoires—oral stories, gestures, and habits passed as embodied memory, two modes of relating to the past that have a complex relationship to power and social difference.[9] Podcasts about the past, at their best, represented an attempt to hybridize these two modes and, even when at their worst, still performed the deep work of turning written words into living sounds, foregrounding replay, reenactment, and revival.

They also seemed to thirst to help listeners know something anew. Some took well-remembered recent history and made it strange on purpose, introducing it through relatively unknown or seemingly discorrelated vignettes. Vann Newkirk II's 2020 podcast *Floodlines*, about 2005's Hurricane Katrina, begins with the story of an enslaved man named Richard, survivor of the famous Last Island hurricane of 1856 who braved the storm in a stable with an old horse, as slavers died around him. Dan Taberski's *9/12* podcast explored how the post-9/11 era evolved by remembering the infamous day as it was experienced by an incredibly isolated group of people with no access to the news besides an emergency satellite phone, because they were filming a BBC reality show that was recreating Captain James Cook's 1768 voyage to Australia, and on the day of the terrorist attacks their ship was becalmed in the shark-infested waters of the Timor Sea. Both narratives use these devices to give listeners an experience of an unknown story that reframes and recontextualizes the experience of well-worn, old information, offering in their inaugural moves a process of re-knowing whereby a main text is approached by way of a haunting image that functions like a manuscript's frontispiece. The process of knowing was also foregrounded whenever a podcast was so thorough as to be nearly manic in its scope, like Tyler Mahan Coe's dives into country music hits on *Cocaine and Rhinestones* (one hour on Merle Haggard's "Okie from Muskogee," two hours on Bobbie Gentry's "Ode to Billie Joe," nearly four hours on Jeannie C. Riley's "Harper Valley P.T.A."), Karina Longworth's epics about Hollywood history on *You Must Remember This* (nineteen episodes fact-checking Kenneth Anger's infamous exposé *Hollywood Babylon*, sixteen episodes on the Hollywood blacklist period, nine episodes on Polly Platt, a prolific Hollywood producer), or Dan Carlin's slogging war narratives on *Hardcore History* that became chart toppers (four hours on the Punic Wars, seven and a half hours on the Mongol Empire, twenty-four hours on World War I).[10]

A thirst for knowing is also evident, paradoxically, in the surprising lack of success that podcasts had at achieving it. In the true crime world, for instance, while a few select podcasts produced exonerations on procedural grounds (some of which are discussed below), few can claim to have actually cracked cases. Indeed, it is startling how "unsolvedness" characterizes the conclusion of many admired podcasts in this period and beyond—*Phoebe's Fall*, *Hanging: Boy in Barn*, *The Ballad of Billy Balls*, *West Cork*, *The Missing Crypto Queen*, *Three Rivers*, *Two Mysteries*, *The Trojan Horse Affair*—and its mood lingers in the negative prefixes of so many podcast titles, such as *Unsolved*, *Unresolved*, and *Unfinished*. In many of these cases, hosts and researchers explore their own struggle to sort between justifiable beliefs and rumors, opinions, and guesses, theatrically exhibiting a kind of agony surrounding the constraints that knowledge faces and a desire to overcome them. This was not only limited to crime podcasts. In a 2017 essay, *Invisibilia* producer Alix Spiegel notes that a typical conclusion to an episode of that program rarely has a "strong authoritative position about what to think," quipping that if her editor would allow it she would end every story by saying, "Beats the shit out of me, your guess is as good as mine."[11]

In many of these cases, it is as if the purpose of the podcast is satisfied not by simply explaining a thing that producers think we don't yet know about but instead by wrestling with *how* to know about that thing in the first place, which is why such a premium was placed on process, something often more vivid than the story itself. *Here Be Monsters*, a podcast produced by Jeff Emtman for KCRW that billed itself as "about the unknown," frequently used experimental sound capture and interview techniques to arrive at new ways of knowing his subjects. In "Eleven Trips to Dreamworld" in 2015, Emtman created a fully soundscaped episode out of dream recollections harvested from tape recorders he distributed to volunteers the previous year. In "Riptides and a Sinking Ship" in 2017, he discussed the fear of drowning, past-life regression, and near-death experiences with a woman who had recently survived a riptide drowning trauma, a podcast recorded with a hydrophone in a bathtub as she repeatedly submerged herself.

To give shape to these many scattered signs of an epistemological turn in podcasting, this chapter focuses on what I call *structures of knowing* in shows of the period. This term is patterned on the well-known concept "structures of feeling" coined by cultural theorist Raymond Williams in several texts in the mid-twentieth century.[12] A mercurial concept, Williams used it to consider the meaning of material cultural performance through signs of how feeling is thought and how thought is felt. Studying a culture's structures of feeling at a given historical moment is unlike employing peer concepts that developed around the same time among Marxist theorists and structuralist thinkers, such as the notion of hegemony (Gramsci), epistemes (Foucault), paradigms (Kuhn), or worldviews (Redfield). At a time when theorists favored narratives of power and knowledge as obdurate as the institutions that perpetuated it, Williams focused on subterranean cultural flows that thrive in everyday behaviors and artistic gestures but are not yet part of institutions or governing languages. A structure of feeling represents social experiences stuck "in solution," like salts trapped in a solvent awaiting a catalyst or current to solidify into crystals; there are typically several such structures vying for prominence in a given setting.[13] As Stuart Middleton has shown, the term emerged in Williams's thought alongside his definition of dramatic "conventions," the socially shared agreements where performers and audiences meet, standards and values that no individual can decide upon but are shared in community.[14] For Williams, structures of feeling don't simply reflect structures of power but represent ideology in a state of emergence, even if a given structure never gels into a dominant paradigm. Williams wrote of the structure of feeling in artwork as something that is lodged not in its explicit material but in its unease, stresses, and latencies, in experiences to which available forms and categories do not speak. These are mixed experiences that we nonetheless perceive to be structured and energized, feelings that are "at once interlocking and in tension." Structures of feeling are often what stick with us about art and its circumstances of reception; reduce a work to just its end state, its winners and losers, and you've hardly experienced the piece at all.[15] Lawrence Grossberg reads the concept as a solution to a tension in Williams's thought that sought other possible ways of being in modernity. He writes, "The structure of feeling is the virtuality, the becoming, of the multiplicity of modernity."[16]

I have always felt that the term "structure of feeling" is a particularly clarifying way of describing narrative radio, a medium structured by rules yet in constant evanescence, embracing a poetics that works in prompts, hesitations, hints, and invitations, as Clive Cazeaux and others have persuasively shown.[17] Radio's forms seem to match feeling with structure and structure with feeling. Jason Loviglio is right that the cliché of the "driveway moment"—when a commuter is so spellbound by a radio story playing in their car radio that they linger in the vehicle even after they have arrived home, an experience that offers a kind of "restorative solitude" in the typical middle-class neoliberal workday in the age of petroleum capitalism—is an excellent example of how radio quite literally structures feeling, a form that NPR has essentially made into its brand.[18] I link my approach to the epistemological turn in podcasting with Williams's idea to highlight this connection, and also because I feel that, like a structure of feeling—an unfinished cultural theory about the presence of unfinished cultural forms—the kinds of knowing that these podcasts embody is never quite articulated or established as a status quo. You can only capture it in details, turns of phrase, compulsions, moods, and reversals. Story podcasts of the period rarely have a stated "philosophy of knowledge" so much as they have habits of how to go about acquiring knowledge and formulas for conveying it to listeners, along with palpable feelings toward what they struggle to learn and uncover, all of which are multiple, overlapping, and in constant evolution. I follow Williams's use of a gerund ("knowing" rather than "knowledge") to help convey the inherently in-process nature of this structure and to avoid the impression that I take the models discussed below as solidified, public, paradigmatic knowledge in any *Order of Things* or "Great Tradition" way, the sorts of approaches that Williams was trying to show to be inadequate.[19] A great deal of knowing happens in podcasting in this period but very little Knowledge in the grand sense.

With this in mind, this chapter looks at three structures of knowing that arose in narrative podcasting around the same time. I call them the familiar, formal, and recessive. The first concerns how structures are created and conveyed with a sense of distinction between those who are familiar with the subject matter of the story or the experience that it accesses—I intend for "familiar" to evoke "family"—and those who are not. Knowing is "structured" in the sense that, while the podcast is open to all who may stream it, it also offers a privileged entryway for those whose experience is most intensified by its narrative, as a result of identity or shared history. Leila Day's "The Birth of Solomon" piece mentioned above is a good example of this, a story that is both by and for Black women (particularly those who experience pregnancy) in a way that is integral to its mode of narration. This is not a mere formal idiosyncrasy of the piece. The episode is a warning to women of color about a danger they face far more acutely than others who might be listening and a story about a kind of grief linked to their intersectional experience, even as it speaks to a wider population in a separate second voice. In this way, podcasts of familiar knowing are marked by a layered, sophisticated understanding of their own "for whomness" that recognizes the political and moral complexity of the very property of address and constituency.

The second structure of knowing highlighted below is tied to narrative design and looks at how knowledge about events is thematized, with podcasters presented as journalistic searchers or activist archivists, ferreting out facts that resist being brought to light. These pieces are often noteworthy for their sheer industriousness, but there is an art to dramatizing that industriousness, too. I describe this structure of knowing as "formal" to indicate that depicted episodes of research are also used as a way of adding emotional modulations to episodes and seasons. Scenes of finding knowledge are woven into the fabric of these podcasts with care, even contributing to their shock effects, cliffhangers, sudden reversals, and moments of alienation. In these examples, I am particularly interested in how impasses, breakthroughs, and revelations are used in strategic locations to shape shows and institute larger rhythms and densities to their narration across and within episodes. They are formal in the sense that they pertain to the "form" of the piece more than the content.

The final structure of knowing that this chapter explores is the most pessimistic, one that focuses on podcasts in which knowledge seems to be constantly slipping away from us, from episode to episode. These narratives of recessive epistemology tend to have international parameters or to take place over long time scales and distances. They share a number of aesthetic features that I explain in detail, features that evoke a world that is hard, if not impossible, for the intellect to penetrate and in this way make for a reversal of the values of recent and historic positivist trends in radiophonic documentary. All three structures of knowing have ways of alerting us to obstacles past which a listener, an investigation, or a narrative cannot go. But narratives of recessive epistemology are particularly noteworthy for promoting an alternative view of the kind of realism to which audio producers found themselves beholden, reaching even further back into the history of audio for an idea of the world that is associated not with recent generations but with early radio, a world that is disordered, contingent, and fraught by static. Attempting to make "something out of nothing," podcasts of recessive epistemology lure listeners with their own guilty desire for more and more nothing.

The American Style and Its Critics

Before digging into the three structures of knowing directly, I would like to back up and explore how an epistemological turn emerged in the first place, the pursuit of a "feeling for knowledge" coming to be a preoccupation of the medium, particularly when it comes to what Siobhán McHugh identifies as "the American style" of storytelling, an approach focusing on "hand-held" or "host-driven" narration guiding listeners through a story in the hope of helping the listener achieve empathy with its subject. The form was part of American storytelling for many years, but it was perfected in the 1990s by the popular *This American Life* technique of formulating radio nonfiction into the form of personal journey that ends in a reflection, joke, or epiphany.[20] In radio, as Biewen and Dilworth have observed, the American style often uses subjectivity to arrive at "something closer to the real" than what "objective" conventional news-based work could offer.[21] The "closeness" of subjectivity is also related to radio's long-standing rhetoric of intimacy, the feeling of traveling alongside a proximal entity as an individual, something that had been a feature of American radio poetic style since the New Deal and that became formalized with the incorporation of NPR in 1970, whose statement of purpose speaks to many of these themes.[22]

The style had well-defined elements that have been described in various ways regularly in books, websites, and seminars by producers associated with it. I cited one such formula as part of my definition of podcast narrative at the beginning of this book, what Alex Blumberg called the "physics" of a story. That physics consists of four elements: a sequence of actions, telling details, a point or punchline, and a moment of reflection.[23] The Chamelecón piece described in the opening of this chapter has all of these features, and Carrier deploys them with painterly control. It meets the American style minimum requirement by virtue of depicting a story through a personal journey of discovery, along with extensive narration, explaining the depicted spaces encountered. It embraces subjectivity, unafraid of framing the story as it meant something to its narrator, who focalizes events. It also has all the "physics": a walk through the town from day to night (sequence), closely observed minutia like the nine-inch speaker at the rap battle (telling detail), the moral about making something out of nothing instead of building walls (point/punchline), and a moment of bringing it back to the mission of the podcast itself (reflection). This style has long acclimatized listeners to its rhythm so that we know what to expect from it and would be dissatisfied if those acquired expectations are not met. Within this flexible framework, there still remained much room for personality; Carrier's laconic idealism gives his work a signature that other practitioners of the style might not have. Audio storytellers taking this rough structure had been making award-winning radio for decades, but it was really in the 2010s that it moved from the world of radio to the world of podcasting, in parallel with the shift of prestige audio.

Inevitably, this set of norms drew criticism when it moved into the podcasting space. Critiques took place in two main phases. The first came in months after *Serial*, during the first wave of true crime shows. These critiques were largely about the journalist's approach to using various elements of the American style—the kind of criticisms peers might raise. One frequent concern was a failure to engage in systemic critique, with many podcasts focusing instead on individual cases, rewarding apolitical approaches by foregrounding close observation of fragments of life and experience, like telling details, punchlines, and epiphanies at the expense of broader trends. Many from the public journalism world recognized this issue and took alternative approaches. Perhaps the best example came in the first season of *In the Dark* by APM Reports. While pursuing the egregiously botched investigation into the murder of Jacob Wetterling in Minnesota, the podcast at the same time looked at the politics of police malpractice and the ethics of police database profiling, thereby turning the focus to a lazy and incompetent system of criminal justice. This approach advanced "the scope of possibility" in true crime, as the Peabody Committee put it, in recognizing the show.[24]

Another concern in this phase was the balance between a need for compelling characters and settings on the one hand and a duty to reflect the obdurate reality of society on the other, suggesting a separation between these two impulses. If a reporter frankly isn't so sure about her sources, should she report at all? Is the use of uncertainty only a lure to encourage listeners to recombine evidence in online forums, something Ellen McCracken has argued is central to the commodity form of the true crime podcast?[25] This critique hits home whenever a podcast fell for a colorful fabulist, as in the 2018 *Caliphate* scandal, for instance.

A final kind of critique had to do with the ethics of focalizing events through a journalist. I think there is a sonic dimension to this concern. To think about perspective in audio, I developed a concept I call audiopositioning, which focuses on proximity at the acoustic level, paying close attention to where we are according to what we hear.[26] In true crime, for example, we "are" usually on a reporter's desk, in her car, and at her home; often she literally has us in the palm of her hand "inside" her portable recorder moving through a space. Our audioposition could be mistaken for one of her everyday handheld possessions, so it is small wonder that we feel manipulated. Even if that were not the case, there is an inherent openness to critique in the genre, as any narrative device that betrays rhetorical emphasis on audioposition immediately calls to critical attention other possible audiopositions that the piece did not elect to take and, in a larger way, constitutes the storyworld of the piece as a dimensional field cathected around points of view rather than one with a single objective nature that is still there when we stop recording. An argument about audioposition is in essence an argument about positionality, with all that term entails, expressing the obvious point that there is no non-audiopositioned storytelling.

The problem of positionality was hardly lost on producers. In his talk at HearSay, Carrier recognized that in the real world, acquiring tape in a place like Chamelecón wasn't really "making something out of nothing" but involved privilege, and incoming donations. His own positionality as an outsider also came through. I couldn't help but note how, although adopting an anticolonial stance, the piece still turned out to be about reflections of America through and through—terrorized by extensions of LA gangs, Hondurans turn to Bronx hip-hop culture, and an American public radio producer sees this as an antidote to racist Fox News "caravan" narratives. The whole piece perceives the Honduran town through a quadrilateral of American cultural forms, left and right, high and low. The story did not have to be framed around Trumpism, after all, but that is nevertheless the "lesson" that Carrier took to his audience, the one he (probably correctly) expected his audience to draw. His comments conceded his role in making the meaning of the piece. "Reality is happening all the time," I remember he said in his talk. "We are the ones who contextualize it." In an essay in John Biewen and Alexa Dilworth's book *Reality Radio* some years earlier, Carrier had expressed this same sentiment while discussing his self-education in audio, learning to focus on what it was about his tape that really resonated with him.[27] It is no surprise that the story is strongly shaped by the argument about art that Carrier wants to make: it's "aboutness" and "interestingness" come to us premixed. An uncharitable reading of the piece might align it with what journalist, activist, and critic Lewis Raven Wallace has recently called an "extractive" approach to journalism, one that treats facts about the lives of subordinated peoples like coal in a mine to be drawn out and refined by and for others, allowing us to tell a story to ourselves about ourselves, one that only passes through its putative subjects, leaving them and their interests largely unrestored.[28]

In retrospect, many of the concerns raised during the first phase of critiques of the American style were more epistemological in nature than they may have first appeared to be. When they seem to question the ethics of creators, for example, critics are actually questioning the structure of shows in how they relate information. It is the serial form, Ryan Engley rightly insists, that is at the root of the "ethical" complaint, in that it amounts to a critique of how information is sequenced, what facts we are prompted to credit and when.[29] Furthermore, in questioning the emphasis on perspective, we sometimes ignore how surprisingly unsure that perspective tends to be. As Loviglio has noted, many of the most exciting episodes of *Criminal* tended to feature not the typical heroic epiphany following a lengthy span of overdone storytelling but "allusive, meandering storylines that prize serendipity over the formal predictability and procedural plotting."[30] As Rebecca Ora has shown, these shows can linger on theories of deception that do not even require fabulists to know they are lying, leaving a chain of unreconstructible events.[31]

This lack of certainty is not always a posture. Many of the first-wave critiques begin with the positivist assumption that we inhabit a highly knowable world, and any failure to engage that knowable world is a sign of ignorance, laziness, or a hidden agenda. But, radio stories are often tales of disillusionment with this positivist stance, a tendency that only grew in the podcast era. Indeed, as I explained in the last chapter, what made *Serial* so innovative was how the journalist at its center investigated her own process; one of the recurring themes of that narrative is how hard it is to remember details. In light of this, it is clear that behind whatever ethical jeopardy reporters courted in "American style" podcasting, there stood epistemological quandaries whose lack of resolution persists even subsequent to any ethical amelioration of perspective or style. And, as Michael Buozis has argued, there is perhaps a surprising politics to highlighting the constructedness of the pursuit of reliable knowledge, in that doing so implicitly critiques the shortcomings of actual criminal investigations.[32] If years of reporting and thousands of hours of writing, rewriting, and editing don't arrive at "knowing," then we have even more reason to look skeptically at how much truth is arrived at in rushed proceedings in notoriously racist criminal justice systems.

If the first wave of criticism of the American style had an undercurrent of concern about knowledge, then outright skepticism emerged in the second wave, which looked at the style itself as a problem rather than its execution. The narrowness, both apparent and hidden, of the method was a topic in many pieces in the year preceding the pandemic, a time in which playful new ways of telling stories began to emerge. Poets led the way. In the experimental piece "A Cow a Day" for *Between the Ears* on the BBC, Pejk Malinovski made a beautifully minimalist sound narrative by following a cow around the holy city of Varanasi on the Ganges River, a boring story about a boring day—but boring in an important way. In *Have You Heard George's Podcast*, several episodes take the form of "Bedtime Story," in which George the Poet's spoken word improvisations create worlds in the minds of half-dreamed characters, blurring the boundary between narrative-driven story and the joy of a well-made rhyme. At times the flow of spoken word follows the story, and at times the story follows the flow. In Ian Chillag's *Everything Is Alive* podcast, meanwhile, the host interviews a can of soda imagining what it will be like to be drunk, a grain of sand trying to understand why humans can't handle silence, and a literature-crazy towel who fantasizes of toweling off Mr. Darcy from *Pride and Prejudice*. The show implicitly proves through humor that our desire to experience empathy requires little more than a construct on the other side to facilitate it.

There were also more directly critical reflections on the meaning of story. On April 4, 2019, the very day that the HearSay Festival kicked off in Kilfinane, KCRW's *Organist* podcast featured an episode entitled "The Narrative Line," in which host Andrew Leland explained his own experience with retinitis pigmentosa, a disease that forces his vision to decline. Learning, after many years, that the narrative he had been told about the disease progression had been incorrect (Leland's vision would decline in a slow, logarithmic linear fashion rather than suddenly fall off a cliff, as he had long believed), Leland reflected on the narrative lines we put ourselves into, exploring alternatives to the podcast formula he heard all around him, with its "imperative to make sense of the world, to put life into a tight narrative arc, a clear sequence of events that's followed by a powerful, relatable moment of reflection" that leads us to "learn something about the world." Just ten days later, the *Invisibilia* podcast released "The End of Empathy," which featured work by Lina Misitzis that took to task the premise of the public radio effort to create stories that generate listener empathy toward a subject, showing how one cut of an interview tape with a young man could result in the story of a pitiable loner with a crazy ex-girlfriend, while another cut of the same tape reveals a dangerous incel. The exercise brought host Hanna Rosin to question the ethics of choosing one edit over the other, asking why it is that public radio as a culture felt it was so important to put empathy with others—irrespective of who or why—at the center of its narrative practice, every time.

Rethinking the reuse of relatively simple and predictable narrative arcs in soundwork was one thing, but a strike at empathy was tantamount to an arrow to the heart of the public radio agenda and aesthetic. Ira Glass has often called radio a "machine for empathy," and the emotion was often linked to "intimacy" as a mechanism by which radio, in the words of Siobhán McHugh, conveyed "the messy contradictions that are real people"—empathy is bound up with radio's purpose in a democracy, its aptitude as a medium, its prevailing philosophy about the world.[33] Virtually no text about modern radio excludes the word "empathy"; most enshrine it. Indeed, it was all over the HearSay Festival program at that very moment. Eric Nuzum, longtime NPR program head and recent podcast evangelist, hosted a talk called the "Unintended Consequences of Empathy," exploring some of the same issues as the *Invisibilia* episode. An arts project called the Empathy Museum sponsored a piece in which attendees visited a shoe shop with 150 pairs of shoes to try on the shoes of Syrian refugees, war veterans, or neurosurgeons and "walk a mile" in them as you listen to their story on headphones.

Still, there were many questions about whose interests "empathy" served and even whether it was a fantasy to begin with. In a profile of the public radio world from around this time, writer Maya Dukmasova pointed out that the kind of empathy offered in radio was often narcissistic, explaining that she found herself weeping while listening to the stories of strangers, without ever having contact with them. "I suspect that rather than empathizing," she wrote, "I'm more often just relating things to my own experiences, moved by a stranger's voice sharing feelings I'm all too familiar with." This, after all, is the point of the exercise. "Every documentary maker wants to move his or her audience," wrote John Biewen, the head of Duke University's Center for Documentary Studies, in a *RadioDoc Review* reflection about his own craft. "Yes, you've got some information you want to pass along, but you work hard at the craft in an attempt to take listeners inside the story, to prompt them to feel something."[34]

By the time of the pandemic, some had lost faith in the post–*This American Life* model for more professional reasons having to do with the swift capture of public radio culture by tech culture. Not only were narrative and empathy no longer commonsensical goals for these critics, but their discursive adoption in the private sector made what had been cultural values in public radio seem more like mere features to expedite content commodification. Producer James T. Green has an essay that speaks to this theme, reflecting on their time learning the style of narration described above at Gimlet in 2018.[35] Green was told that the core of any story was "this is a story about X and it's interesting because Y," explaining that according to their mentors, an example of a "bad" story would be something like the story of an unhoused person who got off the streets thanks to a treatment program, while a "good" story is about the same unhoused person learning surprising life skills from the streets that today they bring to bear on their new job at a mutual fund. It is obvious that the listener who is "surprised" by the latter story probably is presumed to have started out having heard many stories about the unhoused population before, and this one will stand out because it thwarts their assumed expectations of such stories.

But so many already classed and potentially racialized assumptions already underly that calculus. As Green explains, the model "hinged on the idea that a homeless person would only provide lessons that the listener could relate to, only if we know that this person now 'productively' contributes to society, in the financial sector of all things." In other words, the "good" story begins its life boxed in, with plausible versions of its central topic already deemed humdrum and discarded (what if this story was "about" the importance of addiction treatment programs?) or uninteresting (such as the forces leading to the housing crisis in the first place). In this way, the formula is exclusionary, hiding the political process of deciding what a story "is not" behind an affirmative statement, sublimating an interiorized sense of what audiences would find exciting without interrogating what it is about those audiences that suggests as much, while encouraging audio workers to chase success by acculturating to the taste profiles of superiors, leading to homogenization of approach. As Green observed, this model of storytelling also flatters the values of investors, eager for a slide deck–ready presentation about how to "scale empathy."

In light of these waves of challenge to the American style, on both the short and the medium terms, it is perhaps not surprising that creators began casting about for alternative ways to think about stories in an audio medium. The epistemological turn can, to a limited degree, be understood as a way of responding to and managing critiques of the American style, both to improve that style as it is and to imagine what might succeed it. I note that in neither wave of critique of the American style were criticisms based on its beauty or appeal—no one seemed to be saying it was boring, formulaic, or passé—but instead took the form of concerns over how and for whom it conveyed knowledge. And these concerns coincided with the time of #MeToo revelations and hard truths about inclusiveness in the audio world along axes of race, gender, and orientation, lending additional force to suspicions about shibboleths that had guided the American style, including the value of subjective knowledge, the universality of narrative, the imperative to find lessons, and especially the urge to provide empathy. Of course, empathy was by no means rejected as a goal of prestige audio. It remains to this day the main aesthetic effect by which many podcast and radio producers measure their work. But one result of the way the audio world came to terms with itself in the 2010s was a recognition that empathy had become a suspicious fantasy, a cursed optimism, one that could conceal power imbalances and was too easily commercialized. So, while the heart of empathy still beat in the medium, the redoubled pursuit of it was increasingly hard to propose without a deeper discussion of its liabilities alongside its virtues, a development that brought much needed humility to the hitherto vainglorious tone of the approach.

To sum up, while there is no particular cause and effect relationship to point to, it is clear that as the aesthetic of empathy within the larger emotional repertoire of the American style came back down to earth, it found there a fascination with how to know things in the first place. Podcasts began to focus less on connecting presumed upwardly mobile liberal listeners with "others" in order to produce an emotional event for the listener's jouissance and more on finding stories and experiences that were far more difficult to obtain and whose lessons were uneven, not easily explained, unprovided, and at any rate unable to produce relaxing cathartic events. Many podcasters found it unseemly, or even unethical, to impose upon their material idiosyncratic concerns—to be "the one who contextualizes"—particularly when confronting obvious stories of privilege or vulnerability.

In fact, a useful illustration of this recoil can be found in another podcast piece about the crisis in Honduras. By coincidence, the very same year that Carrier traveled to that country to record his story, *Radio Ambulante* producer Daniel Alarcón made a similar journey, spending three weeks in Tegucigalpa and San Pedro Sula, producing a feature entitled "No Country for Young Men" about his experiences. A dimension of this story did deal with providing empathetic engagement in a traditionally radiophonic manner, in much the same way as Carrier's work. Alarcón begins his narrative with Rosa, a Honduran migrant working in New York, and the guilt she feels for leaving behind her daughter Amalia; we hear terrifying tales of how Amalia was nearly killed by a gang member trying to prove himself and other stories of murder and loss that he learns about while traveling in the region. The episode even ends with a defiant song, just like the Carrier hip-hop battle episode. And yet, the *Radio Ambulante* story also recognizes that the real thing it had to convey was the way paranoia seeps into the body when living under dangerous conditions, the way dreams are robbed from Hondurans. It's a kind of visceral experience. "I wanted to tell a *different* story," Alarcón explained, wanting to find some way into the story other than the sheer fear of violence, something that his audience would find unexpected, a typical desire in the creation of post-*TAL* narratives. "And I couldn't: it seemed like there was no other story. Plain and simple. There's no issue more important to Hondurans than this one."[36] Starting out by trying to create a "this is a story about X and it's interesting because Y" narrative, an approach that tilts toward the listener's sensibilities, Alarcón instead decides this is a story about Honduras and focuses on what matters about it to Hondurans, tilting toward the lifeworld of his subject as he understood it. What's more, in doing so, he finds himself focusing on a feeling that he struggles to think through and convey, and this struggle becomes a part of the piece. It's a story that doesn't lend itself to a "point" in the way Carrier leaves us with a passionate vision for making something out of nothing, and this different approach is the source of its merit.

This response, Alarcón verbalizing his recognition that his role in the situation was to express what was important to Hondurans rather than what was impacting to himself or might be unexpected and surprising to his audience, suggests a narrated ethics that the next section considers in greater detail: how questions of "by whomness" and "for whomness" came to the center of how podcasters thought about what listeners need to know in the late 2010s and early 2020s. Throughout this period, podcasters used narratives to find out unknown or long-buried things, to answer questions, or to provide experiences, increasingly with the histories of marginalized peoples in mind. Sometimes this meant making choices about what kinds of experience to bring into the foreground and what kinds to leave in the background, deciding whom different kinds of knowledge was for and even abandoning the pressure to find a large, surprising "point." Over time, through these various efforts, podcasters of this generation came to depict a world that comes across as less stratified, more complex, and less knowable than the one that their predecessors had taken for granted. That development ultimately resulted in a feeling unique to the period, a dystopian mood hovering above a utopian moment.

The Familiar Impasse

Are podcasts getting at the information that matters, are they offering it to audiences in an honest rather than a sensationalist way? Are they taking the positionality of reporter, subject, and audiences into account? Does it matter that few investigative works achieve a resolution of their central mystery? These are some of the questions raised by podcasts that emerged during the epistemological turn. They were also evident in podcast criticism. A particularly cutting example is a 2018 critique by Sarah Larson of the *New Yorker* of the *Last Seen* podcast by Kelly Horan and Jack Rodolico of WBUR radio.[37] The podcast, which deals with a famous unsolved 1990 Boston art heist, is praised in the review for its professional original research and colorful cast of criminals, museum cops, and attorneys but also savagely critiqued on pretty much all the grounds cited in the last section above. The podcast is too juicy in its approach, failing to focus more powerfully on broader questions of the meaning of art in our lives. It is criticized for its use of audioposition, lingering too long in dusty attics, as well as for narrative handholding by Horan, whose voice seems to be "constantly confiding [. . .] a delicious secret, no matter what she is saying," to Larson's ear. It is critiqued on the grounds of standing—many of the interviewees had already written published works on the heist, suggesting there is little new in the podcast itself. There is also a concern over the ethics of conveying information. Larson highlights how the two journalists show audible reluctance to tell investigators about a lead on a Florida site where the paintings may be stashed, lest it ruin the podcast. Larson concludes by mocking the ending, in which "our giddy reporters" fly to Orlando for an "agonizingly drawn-out scene" of the blow-by-blow excavation of the Florida site, after which all they find is an old buried septic tank. "*Last Seen*, like so many podcasts of its kind, having dug and come up without answers," Larson concludes, "is left to poetically invoke life's mysteriousness and the existence of hope." Many podcasts turn out this way, with little to show besides poeticized failure. In this case the absence of closure creates the precondition for critique. Had Horan and Rodolico actually found the artwork, Larson's acerbic takedown of the poetic habits of the American style itself might not have been written.

The overdramatization of *Last Seen*'s last scene, though unsatisfying to Larson, is exemplary of a tendency among podcasters of the 2010s to powerfully dramatize impasses. Many investigated stories have such impasses, often more than one. But not all of them come across as indulgent; in other cases, they are precisely what make stories powerful and important to pursue in the first place. Consider the case of *Somebody*, a 2020 podcast release from Tropic Studios, *The Intercept*, the Invisible Institute, and IHeartRadio, about Shapearl Wells, a Chicagoan trying to understand what led to the death of her son Courtney Copeland, who had been found with a fatal gunshot wound outside a police station in 2016 and died soon after. This Third Coast and Scripps-Howard award-winning podcast is one of a few prominent pieces about crime, justice, and violence that deliberately moved away from the salaciousness that pervaded the genre at the time, adopting an approach that was an alternative to the obsession podcasting discussed in the last chapter. There is no effort to mystify or complicate the story. Quite the opposite; although there are forces arrayed against her, Shapearl is not in a gauzy mist of mismemory and suspicion but rather in a straightforward moral situation, facing obstacles that are tangible. Her son is dead and she deserves to know how and why. At stake is the very worth of Courtney's life, the fact that his death wasn't treated as the death of a "somebody" who matters. The series rests on a logic of affectedness, with Wells's story imposed upon her rather than elected.

The piece has peers in the latter regard. A similar case is Pineapple Street's 2019 podcast *The Clearing*, in which April Belascio pursues the truth about the crimes of serial killer Edward Wayne Edwards, not because she was smitten by the case but because Edwards was her father, and resolving lingering questions about the extent of his crimes fell to her and gave her a chance to confront her father's crimes on her own terms, as she explained in the press.[38] In the CBC's *Finding Cleo*, Christine Cameron is in search of her sister, Cleopatra Semaganis Nicotine, who was adopted off (read: kidnapped) by the government from Little Pine First Nation in Saskatchewan, where they grew up. The host of the show, Connie Walker, herself a First Nation Cree, tells an Indigenous story from an Indigenous perspective, learning early on that Nicotine took her own life at the age of thirteen in New Jersey and focusing the rest of the program on infamous government policies that led to this horrible outcome—shifting, as Neroli Price has pointed out, from the question of "who did it" to the question of why this series of events happened at all.[39] Stacey Copeland and Lauren Knight have written an analysis of the series that highlights the importance of how it explores lineage without the presence of settler voices, instead focusing on Cameron and her brother Johnny Semaganis, emphasizing the "wounded vibrations" produced through generations of violence and cultural genocide.[40]

In all three cases, impasses of knowledge motivate the story at the outset, but they also quickly become only a beginning point in a difficult emotional movement from knowledge to acceptance. All three are stories about grief. Podcasts like *Somebody*, *Finding Cleo*, and *The Clearing* can and were seen to be responses to true crime critiques in the last chapter, as well as to the pulling back from the American style discussed above. These podcasts not only inherently involve systemic critiques of the systems they explore, often in an activist way, but also escape the questions that plagued many of the podcasts discussed in the preceding chapter, by beginning from a place of subjective knowledge rooted in a clearly stated intersectional positionality. These are podcasts by the people they are also about, ones in which women and people of color internal to the situations they are compelled to explore are shown as subjects creating these contexts. Their aesthetic becomes one in which a figure emerges from a ground: a person long trapped by a story told by other people achieves a power over that story by seizing hold of the role of narrating it. Contrast this with the journalist-on-a-journey narrative style, and even the most artful and idealistic practitioners of the previous era begin to seem on an entirely different errand.

Although they sometimes employ elements of the handholding American style, these podcasts are more genealogically rooted in a second stylistic form, the *Radio Diaries*–style approach that emerged from a radio tradition leading back to the work of producers such as Joe Richman and David Isay, even before *This American Life* emerged. The story is well known. In 1993, Isay produced a famous piece entitled *Ghetto Life 101*, a landmark public radio program in which LeAlan Jones and Lloyd Newman, two teenagers of color from the South Side of Chicago, took recorders and produced a piece that walked listeners through their lives. Two years later Joe Richman brought the same idea to a number of other teenagers, producing a memorable collaboration with Josh Cutler, a neurodivergent young man, who agreed to conduct interviews with those around him about his Tourette's syndrome, keeping an audio journal and recording actualities of his life, which generated thirty hours of recordings that the two of them pared down to thirteen minutes. Thereafter the diary format spawned a show and eventually a podcast exploring the possibilities of a workflow in which a subject produces their own story with a producer working as a collaborator rather than as an outside reporter.

In his introduction to a 2016 guidebook to the format, Jay Allison describes it as "first-person tellings of life as lived by the teller" as opposed to reported stories that inevitably insert the reporter's "sensibility and style" into the narrative. Created with a public service mandate and generosity in mind, Allison likens them to "time capsules" that convey the impression of "undiluted essence."[41] Of course, that is only an impression, as Richman explains in some depth in the same guide, since the work of making seemingly non-narrated radio can be painstaking, as it requires sifting through enormous amounts of found sound. But its rhetoric of directness is palpable, and that is part of why this approach had the feeling of solving a problem faced by works like *Serial* and *Last Seen*. Indeed, the personal style comes with its own rigors. As Mia Lindgren has noted, subjective radio of this kind requires a cultivated trust between producers and subjects, and as a result these narratives are subject to just as many ethical pitfalls as journalist-hero narratives are, if not more.[42]

But the approach also has rewards. Reflecting on this style, the radio broadcaster and theorist Sean Street has likened it to a kind of poetry and noted how its focus on everyday speech shifts the work from journalism to oral history.[43] This is an objective of *Radio Diary*–style makers. Richman has written about how to find talkers who speak in visuals, about identifying anecdotes that seem thoughtful without trying to do so, and about the importance of finding poignant accidents. An example he gives is the cough of an attendee at one of Nelson Mandela's trials that is caught on tape, which produced a feeling in Richman that he likens to Roland Barthes's famous notion of a "punctum," that element of an image that seems to pierce or prick us as we are otherwise reading and participating in the image through cultural commitments, codes, and understandings.[44] It is interesting that Richman's definition of a good story has just two elements—action and meaning—and that both pertain to the person being depicted (what happens to them, what it means to them), unlike Blumberg's model—a sequence of actions, telling details, a point or punchline, and a moment of reflection—which pertains to the listener's worldview and experience as the producer understands it.[45] Whatever the true intentions of the producers involved or the ultimate effects, the rhetoric of the *Radio Diary* style is that audio serves the subject, while the rhetoric of the American style is that audio serves the listener.

As the diary format grew more common, it was brought to situations where its stakes were deepened, both in radio and podcasting, often winning awards for doing so. Several diary-style pieces from this era tell stories of migration, such as Leo Hornak's Peabody Award–winning "Abdi and the Golden Ticket" produced in 2015 for *This American Life*, about a Somali refugee's suspense-filled journey navigating bureaucracies and life after winning the immigration lottery, and Sayre Quevedo's "The Return" for *Latino USA*, which won a 2019 Third Coast Award for chronicling the story of poet Javier Zamora, who was forced to return to El Salvador after having his status as a childhood migrant revoked. In the podcasting world, the same was true, with diarist methods brought to high-stakes stories that, when successful, were honored by many institutions: launching *Ear Hustle* in 2017, producers Nigel Poor, Earlonne Woods, and Antwan Williams brought elements of the style to the population of inmates at San Quentin; Honor Eastly's 2018 series *No Feeling Is Final* uses the style as well, with the diarist narrating her own highly personal experience of suicidal ideation and complex battles with finding treatment, a narrative full of metaphor but one that, as Britta Jorgensen observed, tends to abandon verbalization of "big feelings" in favor of music.[46] The objective of the latter series is complicated by a sense of self-strangeness that it also explores in Eastly's inability to even herself really grasp how she comes to survive her own agonies. Jorgensen proposes that it is at the "frontier" of feelings-based podcast narrative. *No Feeling* also illustrates a counterintuitive issue in these pieces, that precisely because diary shows seem so unfiltered, the things that their subjects do not understand about themselves—the puzzle of the self—always stand in stark relief.

In focusing strongly on who these projects were about, it can become easy to miss how complex their work was when it came to who these projects were for. Early examples of the diary approach like those of Isay and Richman seem to presume a certain generous liberal-minded listener on the other side of the radio signal, ready to be transported into the lives of others different from themselves, lives made available to them in an open way devoid of political or experiential asymmetry. But that presumption is not the case in the adaptation of this style to more recent podcasts, which during the epistemological turn grew more pensive about the issue. A moment in the first episode of *Have You Heard George's Podcast?* captures this succinctly. After playing a freestyle by rapper Ambush Buzzworl, narrator George the Poet explains that there are two ways of listening to rap, either for what is there or for what isn't, pointing out that despite exploring the drug game, the song also repeatedly serves as a warning against it. "I can explain that to you because it's my environment," George insists, both acknowledging the "you" to be an outsider and showing how the way the presumed White ear is equally deaf to what he hears as a person of color from a conservative law-and-order justice minister's speeches. Throughout the episode, George is constantly making decisions about whom he is addressing and how, often letting his words be opaque to some and salient to others. Part of the art of the show lies in how its flow identifies hierarchies outside of itself and plays with interpellating identities within those hierarchies.

With this idea in mind, consider the third episode of *Somebody*, entitled "The Police," in which Shapearl Wells focuses on a rumor that, after her son Courtney Copeland was found, he was handcuffed before being brought to the hospital, a story that was alternately denied and then at the same time written off by authorities with the unfounded claim that Copeland had been "combative" with police. Skeptical, based both on her knowledge of her son and on experience of failures of policing in her community, Wells seeks answers from the detectives charged with investigating the case, who prove defensive and badgering, unprofessional both in their transparency about the ongoing investigation and in their manner in communicating with a grieving mother. Learning of the existence of actual video footage of Copeland's arrival at the police station on the night of his death but stonewalled by authorities when it comes to obtaining it, Wells enlists help from a group on the South Side of Chicago called the Invisible Institute, which had been pivotal in the release of the video of the infamous murder of Laquan McDonald by police officer Jason Van Dyke in 2014. The Institute would eventually coproduce the podcast, and in telling this story, like many podcasts, *Somebody* encodes the temporal process of its own creation and distribution into its text. After Shapearl receives the raw footage of the police video (nearly five hundred days after her son died) with the help of the Institute and converts it to a viewable format, we listen to a scene in the episode as Shapearl and her family sit down and watch what happened to her son, with accompanying narration from an external Shapearl in her role as the series narrator.

It is a scene within a scene. With no video for us to see, the playing tape is narrated by the members of the family in the scene of their watching. All we know is what we are told, a situation that creates an acute asymmetry between the people in the show and the listeners, a hierarchy that inverts the typical privilege of the listener as repository of knowledge. In the footage, at first the family group sees what looks like Copeland's car and then his coat on the ground. A police officer looks into the car, and then Shapearl narrates that she sees Courtney on the tape for the first time. The editing brings the music up for a moment, until Shapearl in the scene can't speak any longer and Shapearl the narrator helps her, explaining that Courtney is seen on his knees and reaching up for someone to help him, with no sign of him being combative, as the officials had previously alleged. When the ambulance arrives on the scene in the tape, Shapearl explains he is pulled up forcefully from the ground, but before we hear that, the video goes blurry. This must be the moment Courtney is handcuffed, his mother thinks. The remainder of the scene cuts from the voice-over narration of the events on the tape to the scene in Shapearl's home. Exasperated reactions give way to wordless, frustrated pounding on a table, mic-handling noise, and a wail of pain. It is the kind of scene that can't truly be conveyed in any interview or discussion podcast form. We need not only to hear tape of Shapearl watching the video but to hear voice-over of Shapearl from a later time explaining her reactions, a double scaling of time that narrative formats afford. It also shows in a stark way the awful labor of knowledge that has been put onto Wells in a system that doesn't value Black life. The structure of the podcast has set that labor into a pattern of duration, exposing its frustrations, delays, and silences and framing it in the context of other silences and unknowns in her quest to find out what happened to her son.

At the same time, however, it is clear that the experience of the family is fundamentally different from the experience of the listener, and there is a power to that difference. The police had put up a wall, preventing Shapearl from acquiring knowledge she deserved; overcoming that wall, the podcast shows how the content of tape speaks differently to Shapearl than it can for us, and in this way the podcast reinvents a hierarchy of knowledge that puts its most affected individuals at the top. In this situation, the most knowledge belongs to those for whom it has the most meaning. It is telling that *Somebody*, *Finding Cleo*, and *The Clearing*—all podcasts that successfully reply to many of the critiques of the American style outlined above—are fundamentally family stories. Although we hear what the women at the center of these stories are going through in their searches, we aren't given the luxury to feel we really understand it, that it belongs to its audience as an extracted, refined piece of someone else's life experience. In this way, these pieces also don't presume a listening position characterized by bourgeois White privilege, a listener into whose consciousness all knowledge and experience in the narrative is decanted. Some pieces made in the "radio diaries" style can convey the sense of allowing privileged listeners to pass across what Jennifer Stoever calls the "sonic color line," to experience and learn in an unfettered way about the lives of those outside of their positionality.[47] They seem rhetorically structured as if they are "for" racially unmarked audiences, allowing a comfortable "colorblindness" and an empathy without solidarity. That's not as much the case in this trio of podcasts, however. The scene in Shapearl's kitchen is quite different because it never stops being about and for Shapearl. The key to keeping it that way is a feeling of impasse. We aren't given the belief that we know what watching footage like this is "really" like in an empathetic way. We can be present for something that matters to her, but in the end the *Radio Diary* model is true to its namesake in that, like a diary, it is never really for us, but only for its writer.

This approach is what I call the "familiar" structure of knowing, an approach to podcast storytelling where there remains an area of knowledge that belongs to the participants only, marked as private to their experience of the events and their narration. Rather than setting a godlike listener at the center of all knowing, a being for whom all facts are prepared seemingly complete and whole, these podcasts attempt to set boundaries that put those less affected by the events at a respectful distance from those "in the family" of its circle of meaning, which can be profoundly personal. In *The Clearing*, April Belascio meets with the family members of the people her father had killed and listens to tapes of him; in *Finding Cleo*, the seventh episode has a reunion between Christine and her brother Johnny, whom she hadn't seen in thirty years thanks to the resettlement policy, and a visit to Cleo's grave. In many cases, listeners who share the racialized and gendered experience of these subjects are invited "further into" the work than those who do not, a subtle gesture at ameliorating current and historic inequity.

The familiar structure of knowing could also appear in more conventional narrative formats, too. In the last episode of cold case podcast *Unfinished: Deep South*, a podcast that works well inside the bounds of the American style, hosts Taylor Hom and Neil Shea find themselves at an impasse. For the previous ten episodes, the two journalists had been using interviews and archival research to investigate the 1954 lynching of Isadore Banks, a World War I veteran and prominent Black landowner in Crittenden Country, Arkansas, a crime for which no one was ever charged or prosecuted. After a year of reporting, they had examined Banks's complex relationship to other landowners, his marital indiscretions, the role of racist sheriffs in the juridical history of rural Arkansas, and racial terrorism in the region that took the form of everything from arson to the slaughtering of Black-owned livestock.

Citing the lack of documentation needed to fully pursue a conclusive end to this cold case, and without testimony or reliable archives, the reporters arrived at a conclusion that amounts to guesswork, recognizing that this situation foregrounds "the central role of silence" in maintaining White supremacy and preventing justice. They put forward a theory that Isadore had been killed by farmers in a feud over land—a kind of "critical fabulation," in Saidiya Hartman's sense—meant to stand in for and mark the violence done by Black inaccess to adequate archival testimony.[48] Like *Last Seen*, Hom and Shea come to an impasse, but they find a way to recognize that a politics subtends that impasse, a politics of silence, and they conclude the season not with empty reflections but with a ceremony at the

National Memorial for Peace and Justice marking the victims of lynching in Montgomery, Alabama, where Isadore Banks was honored, his grandchildren attending to mark the day. Hom and Shea's podcast is structured very differently than *Somebody*, but both seek to make the podcast both about and for the people its story affects the most, and part of that is an attempt to make themselves, and the listener, peripheral. In this way, the art of the podcast involves the transformation of an impasse of knowing that deliberately excludes Black life into a system of knowing that centers it.

Formal Structures of Knowing

As is already evident, one striking feature of nonfiction narrative podcasts of the 2010s that made these pieces different from other kinds of audio storytelling is the way so many of these pieces deploy significant narrative resources to dramatize their own challenges. Many podcasts contain specific scenes where a researcher or character faces an impediment as they are trying to know something—it's often a highly "characterized" moment that features a well-conveyed mood of sorrow, danger, tedium, or confusion—and rather than offering the results of that ordeal consequent to overcoming it, the narration really wants to let us in on the details in a highly illustrated way that the structure of the piece organizes for us.

This had been a practice on anthology radio shows in preceding generations—*This American Life*, *Radiolab*, *Studio 360*, *Snap Judgment*—but on those shows, the "story about the story" is often a momentary aside and appears so rarely that these reflexive episodes are among the most scrutinized. For example, *Radiolab*'s 2012 "Yellow Rain" episode features its hosts and producers reflecting on a moment of harm in their segment on the use of chemical weapons on Hmong refugees, in which host Robert Krulwich confronted a refugee on the technicalities of the weapons used against him and his family, insisting on a kind of ownership of the meaning of a story that did not happen to Krulwich.[49] Contrast this moment of reflective contrition with a podcast like *The Trojan Horse Affair*, a full decade later, whose sixth episode shows a moment of breach of journalistic ethics that reveals a professional chasm between an early-career reporter (Hamza Syed) and a veteran reporter (Brian Reed), from which both will learn something, an unusual move in a journalistic text, even if it is an entirely predictable one according to the show's subtext of a buddy-cop routine in which a seasoned veteran jockeys with a rookie, the pair separated by age, culture, and race. In part because podcasts are much longer in total duration, and in part because their most famous precedents create the expectation of "behind the scenes" details on how reporters gathered their material, they often explore the hardest parts of their production processes of discovering information in an exorbitant and structurally significant way. I call these "formal" structures of knowing to distinguish them from the familiar ones discussed in the last chapter and to emphasize that they work largely because of where they fit in a narrative structure.

Perhaps the most prominent example of a formal structure of knowing takes place in *S-Town*, Reed's highly successful *Serial* spinoff that explores the life of the depressed autodidact John B. McLemore, who had brought Reed to his home in Bibb County, Alabama, under false pretenses, luring him with talk of dark dealings involving rape by lawmen and a covered-up murder by the scion of a local magnate. When little evidence of these lurid crimes turns up, for some time it is not clear what the story is going to really be about. Two episodes drift by with many sharp details and amusements, and then suddenly we learn that McLemore committed suicide, a shocking event that deeply affects Reed. The next five episodes become a passion project to take full measure of McLemore's loves and hatreds for his town, his closet homosexuality, his masochistic addiction to tattooing, his mysterious finances, and his possible toxification through exotic clock restoration techniques, all to try to understand his surprising death. Perhaps the most unusual choice from a narrative structure perspective was how the team chose to narrate McLemore's death at the end of the second episode; a more conventional approach might have chosen the beginning of the series or the end. On a repeat listen, an answer becomes clear, as a trace of each one of the themes of the latter five episodes is already seeded in the first two: a set of connected controlling metaphors (McLemore's maze, his clocks, his love of "A Rose for Emily," the Faulkner story that is also the show's theme song); scenes of him playing with potassium cyanide "like a mad scientist"; clues that John is a fabulist by his insistence on local corruption but disinterest in doing actual research to ferret it out (Reed remarks, "John doesn't seem to care that we're not making much progress"); offhand references to his sexuality; hinted rumors about John's secret wealth by his cronies at his hangout bar; hatred of police; and fixation on tattooing and climate change. The podcast deploys all its core questions and themes casually in just two episodes and then provides a shock that makes us forget that it has done so. Afterward, it proceeds to reengage its questions and themes in a methodical way that explores each of them, even as they feel like echoes of forgotten memories. In this way, a reportorial dead end is repurposed as a structuring mechanism. John's surprise suicide at the end of episode 2 is like a prism, refracting each piece of the first part into a rainbow of distinct, individuated narrative threads in the second.

A similar use of impasse as structure can happen in the context of a single episode. Consider the second season of *In the Dark*, in which the investigative team of the podcast looked into the case of death row inmate Curtis Flowers for an alleged mass murder at the Tardy Furniture store in Winona, Mississippi, a crime for which he had been tried six times by the same district attorney, Doug Evans, and served some twenty years. The widely acclaimed season is best known for helping to spur the overturning at the Supreme Court of Flowers's most recent conviction, on the basis of racial bias and procedural failures on the part of prosecutor Evans. The podcast is surely one of the reasons Flowers went free in 2019. Based on the Supreme Court filings, two aspects of the podcast's legwork seem to have directly contributed to this judicial result, and both aspects appear in the series in episodes that have what I call "double impasse" structures that show first a mind-numbingly slow pursuit of knowledge, followed by a nail-bitingly suspenseful one.

The first double impasse takes place in the eighth episode. After an opening act in which host Madeleine Baran explores prosecutor Doug Evans's youth and rise in the context of several key flashpoints in the civil rights era in Mississippi, Baran ends the act just before the first break, describing her research into Evans's twenty-five years of cases in newspaper files kept at the state archives in Jackson. Noticing several news stories qualifying the Flowers verdicts by the race of its jurors (i.e., "The all-white jury found Flowers guilty"), Baran wonders aloud how often Evans used peremptory jury member strikes, a privilege afforded to attorneys prior to trial, to create all-White juries over the course of his long career. This is a kind of information that no agency tracks. The first act of the show then ends with a promise of what's to come ("It took so long to find the answer to that question"). When we return from the break, we learn of a herculean work of primary archival research that takes up the second act. APM reporter Parker Yesko moves to Winona and begins scanning the records of eight courthouses, looking at every criminal case in the circuit since 1992, ultimately digitizing 115,000 pages of documents from local clerk docket books and transcripts of jury selection processes, sometimes stored in old restrooms or jail cells, yielding a total of 418 jury trial records and 225 total trials for which racial information of jury pools was available. Then data reporter Will Craft built a database analyzing the strikes that Evans's office brought to Black candidates and discovered that from 1992 to 2017 the office struck Black jurors at a rate four and a half times that of White jurors—fully half of all Black jurors were struck from juries, as compared to only 11 percent of Whites. Then, in the third act of the episode, Baran and Yesko manage, by dint of perseverance, to actually see Doug Evans in his office, recording a brief eleven-minute exchange with him in the hallway, which is the only time he speaks with them on tape. It is a powerfully emplaced moment, claustrophobic with sonic closeness, denials, recusals, and Baran's comments recorded later unpacking the attorney's stonewalling and vague accusations whenever he mentions how witnesses have retracted their stories. We can't help but wish Baran would bring up the jury trial research that her team has just explained, but she doesn't get a chance before Evans is called away and the episode ends.

In the tenth episode, remarkably, this very same pattern repeats itself. This time the first act focuses on his history of Doug Evans's Brady violations, cases in which the district attorney's office failed to provide defense lawyers with potentially exculpatory evidence. After showing Evans's history of bad practices in this area—none of which had resulted in any meaningful reprimand—Baran again wonders aloud if such violations occurred in the Flowers case, learning from a retired officer that at least one report was missing from discovery and recalling a mysterious page in the file indicating that someone named "Willie James Hemphill" had waved Miranda rights. As the first break comes, Baran tells us it took a year to find what that mysterious page was really about. By the second act we are with the team in an abandoned plastics factory, where arrest records were dumped in one hundred two-foot-tall heaps on the ground, covered in black mold and mouse droppings, near old hospital beds and stacks of rotting boxes. Days of photographing eight thousand arrest cards in an old filing cabinet at last yield a jail record for Hemphill. In the third act, it takes months to track down "Willie James Hemphill" (they found half a dozen from Mississippi to Chicago), until we arrive at the right one, who the reporters manage to find in court on a day he is being arraigned on a marijuana charge. Hemphill confirms that he was a suspect and that he owned a pair of shoes like those linked to the murder, all information that the prosecutor's office had neglected to produce to the defense in discovery, a clear violation of trial law. In a tantalizing colloquy Hemphill denies involvement in the crime, and the episode ends without quite expressing the obvious question in the listener's mind—is this the real murderer? It's a question neither the episode nor the podcast can answer.

Both of these episodes enact the same formal structure of knowing, which has three stages that coincide with a roughly three-act template. First, there is an opening act that sets the theme of the episode (the first about Doug Evans's childhood; the second about his history of Brady violations), culminating in a daunting archival research question before the first break that seems like an impossible task (the ratio of strikes of Black jurors versus those of White jurors; the tracking down of a single name on a single piece of paper in the filings) characterized by a plodding duration. Second, once we return from the break, Baran and her team have ingeniously come up with a way of getting through the archival problem and walk us through its steps and results (the first with legwork and databasing; the second with efforts at the abandoned factory), a process that showcases data labor. Finally, once this is completed, a promising in-person encounter with a live source (Evans; Hemphill) is dramatized, a tense scene that takes an opposite temporality; we have just a few precious minutes to learn something, after having taken a year lost in stacks of paper. The brief drama of sudden suspense nearly fully connects with the slow drama of archival drudgery just solved, but not quite. An archival impasse is defeated triumphantly, a human one persists.

Both *S-Town* and *In the Dark* do resemble earlier forms of podcast and radio narration. Both are still about empathy in a traditional public radio sense, trying to prompt feeling for an individual caught in systems of structural inequality and othering, processes that these podcasts try to understand and ameliorate. But critically approaching these pieces only in terms of the production of empathetic effects cannot really explain their chief formal properties or the way they sequence information. As in many podcasts of the epistemological turn era, in these cases an emotional dimension is less prominent than a noetic dimension. This also helps explain the unfinishedness with which both pieces come to rest. Neither the McLemore suicide nor the Tardy Furniture store massacre ends up being truly "explained" in these stories. There are only theories, hypotheses, and insinuations left out there, and so it is hard to think about these pieces in terms of achieving closure or arriving at an epiphany. But what is definitely "resolved" is the logic of the formal knowledge system that each one adopts. In *S-Town*, by the conclusion of the series, we have explored the full spectrum of material that the first two episodes truncate and the last five expand; in these *In The Dark* episodes, the themes introduced at the beginning are made whole by the end through research, with the loop between the two tantalizingly near closure. In this way, the use of a formal structure of knowing could allow a piece to be dense in its account of the complexity of mysteries it cannot really solve on the surface while underneath maintaining a deep structure that conveys the feeling of a closed loop. We don't know it all, and some things we only half know, but we are made to feel that we know the rules of how to understand the worlds these reporters explore, that we have become worldly enough to recognize how limited our capacity to prove things really is, particularly in a society where knowability is itself implicated in systematic forms of prejudice and oppression. In podcasts that "formalize" their structure of knowing, the reward of acquired empathy is surpassed by the reward of a newly acquired worldly savviness.

In the case of *S-Town* and *In The Dark*, it is clear how to "look for" formal structures of knowing: in the first case by paying attention to a sudden shock and in the second by looking at a repetition of a deep structure. Another way of doing so is to listen for moments when a podcast narrative shifts its own established rules of enunciation. An interesting case takes place in a 2016 episode of *Love + Radio* entitled "Doing the No-No," which explores the thought and practice of bio artist Adam Zaretsky, a particularly cunning episode of this consistently boundary-challenging podcast. The piece starts out as an edited monologue by Zaretsky, a self-styled "Lenny Bruce of Bio Art." Inspired by the BDSM community, queer innovators, and shock artists such as Karen Finley and Robert Mapplethorpe, Zaretsky explains his fascination with reproductive technologies as deviant art. Seeing scientists creating frogs with eyes on their backs that concretize the dreams of surrealists of the past, Zaretsky fantasizes about creating transgenic humans as art projects: babies with antlers, pinstripes, pig noses, or goth-inspired body modifications—the development of transgenic humans inspired by Picasso, Pollock, or Sherman. His account of this practice oscillates between a critique of biopower and an enactment of it, between thoughtful engagement with the social meaning of CRISPR-based gene-editing technologies and a prurient interest in their results. Realizing that human gene editing has largely been discussed in neofascist terms that attempt to perfect the human body, Zaretsky both critiques that paradigm and champions the technology as a way of queering the body. This work takes the form of challenging students to decide if they are willing to microinject plasmids into pheasant embryos, electroporating his own sperm with centipede DNA and arguing that intervening in genomes can help "let the radical porousness of reality flow into anatomy" in taboo practices that he calls "doing the no-no." In the end he appeals to the public, perhaps only as a sinister joke, for those willing to volunteer their eggs and sperm for him to engineer "sculptural" children. "I'm really good at giving people hands-on experience understanding the issues involved with doing the no-no," Zaretsky says, "but I'm also really involved with getting them to do it, coercing them a little bit."

But the taboo at the heart of "Doing the No-No," it turns out, is not about bio art at all. About forty-five minutes into the story, the surface narrative is broken, as producer and interviewer Britt Wray—almost entirely silent to this point—speaks up from the other end of Zaretsky's monologue, alluding to an incident at an artists' event (the name is redacted) that led to his work being banned. She mentions he might want to talk about it and then directly asks him to speak about his shared history with her. There is an uncomfortable silence and then another. The for-whomness and the about-whomness of the episode shift like the sand in a flipped hourglass. We soon learn that the "no-no" that the podcast is really interested in exploring concerns an incident during one of his harebrained experiment-performances at the artists' retreat in which a drunken Zaretsky had pressured Wray into donating biopsied material from her own skin, twice (not just a "no" but a "no-no"), for a project in which he was combining DNA from a number of biological sources into ink for a tattoo. Exploring this incident has been the idea of the podcast story all along, it seems, and Wray's agenda has been not to platform Zaretsky's art theory but instead to ascertain the relationship between his ideas, in which he views bio art as a kind of pornography, and his inappropriate pressure of her and others in pursuing it. It is as if Wray alters the genetic code of the story into something else, putting its puffed mythology under another lens, where it withers. Zaretsky stammers and apologizes his way through the rest of the interview, his revolutionary ideas all of a sudden seeming smaller, more feeble. This reversal makes the story a classic piece in podcasting's epistemological turn, one that employs a Brechtian approach to a formal structure of knowing, luring us into learning one narrativized world of understandings only to switch it by means of a sudden estrangement into a second one that was present all along, hidden underneath, while at the same time revealing the first structure's artificiality and constructedness, and indeed that of podcast interviewing itself.

Long Shots

So far I have largely focused on systems of knowledge and constraint that take place at the beginning of a series and are set aside or overcome for acceptance (shows like *The Clearing* and *Missing and Murdered*), those that take place in a specific scene to ameliorate one asymmetry of knowing by establishing another (*Somebody*), at a key moment of a podcast's overall run (*S-town*), or at particular moments within episodes that punctuate them (*In the Dark*) and institute rhythms of expectation and discovery or undermine them ("Doing the No-No"). Of course, the most common kind of impasse is none of these, but one that comes toward the end of a season, at which point a journalist or researcher concedes they may simply not be able to find the answer to whatever puzzle has preoccupied them. There are versions of this moment that seem to aim for poignance once this juncture is reached (the ceremony honoring Isadore Banks in *Unfinished*) and those that can seem as empty as an old septic tank (the conclusion of *Last Scene*), but no matter what, the narrator has to deliver a valediction confessing the bad news that, despite all efforts, a solution does not seem possible and then has to find some way to bring grace and equilibrium to the process of knowing that has brought us to this juncture. To use critic James Phelan's terminology, these narratives achieve "closure" with a low degree of "completeness," in that a narrative still sustains instabilities within the story—who did what to whom and why—as well as tensions between the narrator and the audience, including the call for more information from the public, who may hear the podcast years or decades into the future.[50]

An interesting case of this is the 2016 first season of the CBC podcast *Someone Knows Something*, which explores the 1972 disappearance of five-year-old Adrien McNaughton near Holmes Lake outside the town of Calabogie in Ontario, Canada. On a fishing day trip with his family, Adrien got bored, wandered back toward the family car, and then was never seen again. Authorities have no idea if he was kidnapped, got lost in the woods, or drowned. Veteran cold case reporter David Ridgen, who grew up in the area, interviews Adrien's family members and others over the course of several episodes, chasing down a wide array of leads ranging from 1950s cars spotted in the area back in the 1970s to the use of dogs trained to smell for cadavers and several dives into the lake. In the eleventh episode of the season, after dozens of interviews and fruitless forays, McNaughton's brother Lee speaks with Ridgen on whether the podcast has been worth it:

> RIDGEN: In finding new information to try to fill the void of "not knowing," we at *Someone Knows Something* shaped our own story out of the mystery of Adrien's disappearance.

LEE: I think our very first conversation I had with you, I worried what was your take going to be? [. . .] You know, did you have conclusions that you were planning on? But I think you let the story tell itself in that way. I think the only disappointment I've had listening to it is not listening to it as much as . . . is it actually having an impact. Is it actually . . . you know, the title is *Someone Knows Something*, but what if no one knows anything?

Ironically, an audible edit rests at the moment Lee praises the show for letting the story "tell itself." Ridgen's negation of his own show's premise is likewise rhetorical. Note how the dispute about his subjective approach is sidestepped at the prospect that the void at the center of the story may never fill. But the passage says something more than even this. Here, the prospect of no one knowing what happened to Adrien may not have anything to do with the problematic aspects of the American style of podcasting or anything to do with structural social inequality or police incompetence. Lee is concerned about something else: that no living person actually knows what happened to Adrien and that the impasse of knowledge can be neither abandoned nor surpassed. He likens this feeling with his tendency to stare at the stars at night and find a kind of odd comfort in its vast unknowability.

This first season of *Someone Knows Something* has several affinities with two other forms of epistemology that I have discussed so far, each of which structures how many podcasts of the 2010s felt about knowing. It is "formal" in its use of a research impasse at key temporal junctures (e.g., when Ridgen plans a dive to search the lake, he must wait a few episodes for spring thaw), and it is also "familiar" in its attempt to transform a failed whodunit into a family story of grief and acceptance. The last episode ends ingloriously with a gathering of the family outside the police headquarters in the hope of reopening the case, only to learn that they needed to meet with a different office; the mistake nevertheless gives the family a moment to grieve. But *Someone Knows Something* also belongs to a crime subgenre I call "stories of recessive epistemology," an approach in which not only is "knowledge" elusive, but the very ability of producers and listeners to accumulate and assess knowledge sinks into a fraught auditory backdrop over and over again across the span of its episodes.[51] To explain this aesthetic and its appeal as a third structure of knowing, I would like to group *Someone Knows Something* with two other podcasts, both from 2018: *Lost in Larrimah* by *The Australian* newspaper and *Death in Ice Valley* by the

Norwegian Broadcasting Corporation in collaboration with BBC World Service. Despite emerging from different nations and production infrastructures, these three podcasts share kindred resonance around recessive epistemology.

Like *Someone Knows Something*, in which the reporter is quite literally wading in a muddy lake looking for someone lost back in 1972, the other two podcasts are also long shots. *Lost in Larrimah*, produced by Kylie Stevenson and Caroline Graham, examines the disappearance in December 2017 of retired larrikin Paddy Moriarty in Larrimah, a tiny off-the-highway Australian town with only twelve residents (nearly all in their seventies), a colorful bar, and a novelty zoo. Why did Moriarty vanish without a trace—was it a sinkhole, an old vendetta, a killer croc? What starts as a locked-room mystery in the open wilderness becomes an ethnography of outback culture in decline, a miniature work of oral history that takes the form of a compendium of barroom folktales of disputes between camel meat pie sellers, grudges over stolen chocolate bars, and insults traded in Fran's Devonshire Tea House. In *Death in Ice Valley*, journalists Marit Higraff and Neil McCarthy follow the cold trail of a mysterious unidentified foreigner known as the "Isdal Woman," a Jane Doe whose burned body was discovered in a valley near Bergen, Norway, in 1970. Intriguingly, all she left behind was a series of disguises, a handful of barbiturate pills, and seven separate sets of forged identity papers. Episodes follow files of the secret police, mitochondrial DNA, isotope tracing, and dim eyewitness memories in rooming houses and fishing villages to try to discover who she was. Of the three podcasts, *Ice Valley* generated the most dedicated following, with some twenty thousand online sleuth fans compelled to work on the case, a community that the narration constantly asks us to join. As of the time of writing, all three podcasts have failed to solve their mystery. Indeed, they have failed to remove doubt that a crime even occurred. If critic Mark Seltzer is right, that true crime is about recombination across a social field, the rejoining of bodies and intentions that have come apart, then unsolved true crimes confess the truth that sometimes neither bodies nor stories nor social fields fit.[52] The irreparable fracture of narrative acts as a kind of metaphor for a fractured and atomized civil society, what Seltzer likes to call the "pathological" public sphere.

To underscore how unusual the trend of such programs is, it is helpful to situate them in the longer history of documentary crime audio, where it has historically been highly unusual for focalizing characters to linger in a state of doubt so acute they are unable to bring order to the storyworld and where true crime has long been a vehicle for the formation of a coherent social space. In the 1930s the first crime docudramas were invented for the purpose of investing the medium with authority and removing public fear of doubt when it came to public order. True crime re-creation shows like *Police Headquarters*, *Gangbusters*, and *Calling All Cars* emerged out of a desire on the part of several actors—police forces, radio sponsors, the networks themselves—to improve their public image in an era when they all faced calls for reform. As Kathleen Battles has explained, this confluence of factors took form by replacing an older, more popular, working-class notion of "truth" with a newer idea of "facts" that valorized police and allowed networks and ad men to be seen as public servants.[53] The insistence that dramatized crime stories were based on "the facts" survived as a convention into 1940s and 1950s programs like *The FBI in Peace and War*, *This Is Your FBI*, and *Dragnet*, programs that nearly always claimed to be based on authentic case files as they worked as mouthpieces for law enforcement. It was the age of the myth of the positivist G-Man, an affectless White male as possessor of truth. As Battles puts it, "Factualness allowed listeners to enter the inside world of police knowledge, but only for the police to speak at its audience. Facts were the nonsentimental, masculine, and scientific counterparts of the truth."[54] And entering the world was the whole point. In the 1930s, docucrime seldom involved mysteries, as the criminal's identity was usually known from the beginning, and the story instead focused on apprehension and the willingness of the public to participate in achieving it.[55] This idea of policing persisted well into the television era. As Claudia Calhoun has shown, it also informs public ideas about police work to this day.[56]

If classic docucrime relied on facts to rehabilitate police and worked to designate them as a mediator between reality and the listener, more recent styles adopt narrative framing to fuse factuality and its interpretation. An act in *This American Life*'s 2010 episode "The Right to Remain Silent" is a good example. In the episode, host Ira Glass spends four minutes leading up to the playing of a tape in which an NYPD officer is caught relaying illegal orders from his superior that beat cops must hand out a quota of tickets in order to keep up revenue flow in the district. The tape itself, from June 2008, is brief: at a precinct meeting the sergeant says, "He [the executive officer] wants at least three seat belts, one cell phone, and eleven others." But to make this piece of audio into a damning "fact" for the audience, Glass has to first testify that it has already caused a local scandal and then introduce *Village Voice* reporter Graham Rayman, who obtained the tape from whistleblowing officer Adrian Schoolcraft and who vouches for Schoolcraft by citing his family devotion to the police force and his "extremely earnest" demeanor. This ritual of vouchsafing integrity is still not enough. Over the next three minutes we hear Schoolcraft describe his practice of taping, as Glass gives us specifics of the gear he used (a recorder the size of a pack of gum—such details lend authenticity) and then narrates Schoolcraft's tapes from at least four different roll call speeches at the precinct (some of which actually take place after the June tape), asking the officer about euphemisms ("the rent's due" means they need money), while providing asides that describe the location and explain rules governing what officers can ask for. Only then do we hear the June tape. Listening, it is actually quite hard to catch the fact that we do not really hear the executive officer utter the illegal order that is the basis of the scandal. We only hear about the statement from a sergeant on a tape, framed by talk between Glass and Schoolcraft, framed by more narration by Glass and Rayman, framed by yet more narration by Glass. My point is that even utterly damning evidence conveying knowledge about perfectly obvious corruption arrives in our ears with an extraordinary amount of preprocessing in this particular narrative idiom. We are not asked to interpret evidence being set out for our ears. That's been done for us, and done so exorbitantly that we cannot help but be persuaded by it.

The old true police shows are ultimately about power. *This American Life* is, by contrast, a show about reflection (another kind of power, really), and to provide this it surrounds raw tape with contextualization that depicts and interprets the world in a single gesture. What both strategies share is a facility for turning audio into a fact we can understand, with little doubt left as remainder. We are not here to wonder what is true and what isn't, if it matters or not, who is left out; of course, we *can* wonder about all that, but the piece doesn't urge that errand. These are orderly moral worlds in which everyone knows how to know and what to know and eventually knows everything. It is against this "positivist" approach to epistemology in radio, one in which all the "knowing" is being done for you so that all the "feeling" can be optimized, that a far more pessimistic aesthetic, one corresponding to an alternative philosophy of realism, could grow.

Elements of Recessive Epistemology

Let me return to the three "recessive epistemology" podcasts with which I started this line of thinking—*Someone Knows Something*, *Lost in Larrimah*, and *Death in Ice Valley*—to highlight some of their overlapping features. If shows from recent radio put facts and knowledge in the foreground, these do the opposite, making it feel as if access to certain knowledge is receding. I believe there are seven interlocking aesthetic features that podcasts of recessive epistemology tended to exhibit in this period, and below I explore them in the form of a list.

1. Glitch Sound Elements. All three podcasts have *musique concrète* design elements in common, among other more complex approaches to sound design. Some episodes of *Larrimah* use a distorted bass warble; others introduce sound effects into music in ways that work as miniature dramatic embellishments. When Paddy's dog Kellie is mentioned, for instance, we hear dog breathing in the underscore; the sound of train cars comes when we hear of Larrimah's glory days before the town's decline. *Ice Valley* often features a musical sting of a muffled woman's voice speaking in an unclear language oscillating across the stereo image, a glitchy musical motif that links scenes. She "speaks" in between lines of dialogue whenever evidence seems promising, as if she is about to finally communicate with us but is held back by glitches in the communication system. There is even a second ghost voice whenever the Isdal Woman's meetings with men are mentioned, as if there is a dramatization taking place a little far off in the back of the audio and we can barely catch it. As Caleb Kelly has explained, in sound art, glitches often suggest both a mastery of technology as well as a crack in the code that opens up into something unpredictable.[57] In this way, the Isdal Woman glitch conveys in miniature the quicksand-like epistemological mood of these podcasts, in which attempts to reach a sonic order only put it out of reach. It makes the podcast seem more sophisticated, but it makes its central mystery seem more opaque.
2. Structuring "Unclues." These podcasts have, roughly speaking, three types of clues. The first is forensic, the sort of details you might find in a typical crime TV show: mitochondrial DNA and isotope analysis that put the Isdal Woman's age at much older than expected and her birthplace near Nuremberg; bank records showing that alcoholic Paddy hadn't bought a beer since the day of his disappearance from Larrimah. A second type is

"possible" clues whose potential meaning is swiftly disposed of: a sock pulled by divers in Holmes Lake, where young Adrien disappeared; a steel spoon in the Isdal Woman's effects. Neither of these tantalizing objects leads anywhere, but they importantly provide more tangents to follow, making the story seem larger and also more tedious, requiring of industry. But there is also a third category of clues that invite unstructured speculation that lasts over the course of several episodes. When specially trained cadaver-seeking dogs present indicating behaviors at specific spots at Holmes Lake, for example, it leads to hours of speculative thinking. We have to wait for months for the ice to break in the time frame of the reporting, and then we wonder, what if the wind were different? Then when divers find nothing, we think, what if we tried again? Several episodes of *Ice Valley* explore handwriting analysis, down to infinitesimal details, such as what makes the Isdal Woman's T-letter formation look like she learned cursive in France and the last, oddly lettered hotel card she signed, by which it seems she is "disguising her disguise." These pieces of evidence invite what Carlo Ginzburg called a "conjectural paradigm" that uses intuition to decode subtle clues betraying identity.[58] But instead of objects of final diagnosis, descended from a venatic and medical paradigm of knowledge acquisition that turns the world into a text to read, clues instead work like refrains. They vanish from the narrative focus and then return in later episodes, refined, only to be complicated again and come around once more. While seeming to generate the affective qualities of what Ricoeur called a "hermeneutics of suspicion," these pieces at the same time keep the pleasure that this reading technique—which interrogates details to yield their hidden meanings, destroys to create—offers from taking place.[59] By the end of each series, we don't know what most of the evidence means or even if we are right to see it as "evidence" in the first place. The loop of conjecture never ends. Perhaps the true function of these "unclues" is to captivate fan communities by providing objects around which to rearticulate but not resolve hypotheses, recursively. They suggest that the podcasts are cognitively structured around the fan subreddit topics they might spawn, showing how deeply the logic of social media communicative patterns is embedded in podcast aesthetics. Podcasts of recessive epistemology even resemble conspiracy narratives of the period, in that they create their own constellation of objects, theories, and details that become a lingua franca to initiated fan-participants and don't effectively signify outside of their circle.

3. Predigital Entropy. True crime is known for deceptive suspects and corrupt authorities, but these three shows don't have real suspects, and police seem aloof or at any rate separate from the story being told. What producers work against is not the dismissal of the value of marginalized lives, and the systemic prejudices these dismissals embody, but forgetfulness and incomplete information. Half the people in Larrimah can't exactly remember on which day Paddy went missing. Ridgen spends several episodes trying to track down a black and white 1956 Dodge seen in the area in 1972, but some say it may have been all black or perhaps blue and white. In *Ice Valley*, several witnesses in Bergen—a waitress, a shopworker, a housekeeper—say they saw the Isdal Woman with several different men, but there is no account of it in the police file. All three podcasts contend with entropy and neurodegeneration, with key witnesses aging and dying, and in a larger sense elegiacally reflect on the ephemerality and lingering opacity of lives that were lived before the arrival of the digital storage age. With all the touted affordances of digital memory, the world remains irreducibly "made" of human memories. It is significant that these stories tend to be told about White subjects in search of other White subjects and that the impeding elements of loss that prevent resolution are "natural" rather than social or political, as was the case for the Isadore Banks podcast, for instance.
4. Surrogate Embodiment. These pieces place their hosts in the same locations as those occupied by missing people in their narratives so often that you could call them elaborate sound walks in the shoes of the disappeared. As a consequence, they also have a phenomenological valence, in which recovery of lost meaning begins with co-situatedness with the lost object. The reporters McCarthy and Higraff follow the Isdal Woman's tracks to prewar Nuremberg, through hotels she may have stayed at and places to which she might have fled. David Ridgen returns to Holmes Lake six times, often walking where Adrien walked. It's like a séance in reverse, in that rather than bringing the ghost

to us, we try to come to them. And when these hosts take on the audioposition of the lost, or even just their barstool, we ourselves stand inside the earspace of the missing, as an addressee, hearing what they might have heard and where they heard it. The breaking of the spell can be jarring in these cases. When Ridgen, walking in Adrien's shoes, calls to Adrien across the lake on our behalf, he is calling to us as listeners, too.

5. A Preference for Periphery. All three podcasts are set on the edge of geographies and international systems. Larrimah sits as terminus of a defunct railroad; the next town down is already a ghost town. The Isdal Woman's movements intersect with the development of an anti-ship missile called the Penguin, a strategic weapon at the front line of Cold War geopolitics, though situated in the remote North. All three podcasts use natural sound to mark episode perimeters—Norwegian rain, Australian night bugs, Canadian lake ice cracking. This affects our confidence in expecting a certain conclusion to the mystery. Even the very recent Larrimah case is presented as unsolvable due to the remoteness of its landscape, which we are repeatedly told is populated by stray dogs and pigs, hawks, eagles, and death adders. The reporter Kylie Stevenson explains: "The police search area was massive, 85 square kilometers, but this landscape is so vast, the grass is so tall, and the scrub is so thick. You only have to move a few meters away from something and it disappears. People in Larrimah don't think Paddy's body will ever be recovered." Like human memory, these hinterlands are themselves slipping away; climate change is an unspoken theme in these warming, flooding settings. You might think of them as eco-sonic works in the sense that they represent a world that is not entirely for us, or our stories, but someplace alien and indifferent to human narratives.[60] The washing away of life and narrative has the feeling of a harbinger of the growing climate crisis.
6. Soundscape-Forward Mixing. It is remarkable how prominent sonic backgrounds are in the mixes of these pieces. In *Larrimah* many events and conversations take place far off mic, and scenes are sonically dominated by noisy refrigerators, trucks, and insects. Compare spectrograms of these three podcasts with typical voice-forward journalistic podcasts and you will see all

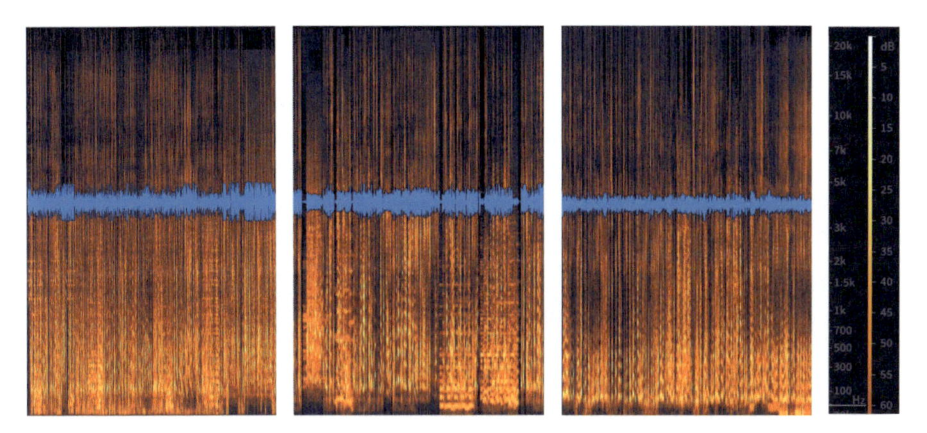

Figure 7. Spectrograms of the first few minutes of *Serial* (season one), *In the Dark* (season one), and *This American Life*'s "The Right to Remain Silent" episode. Decoded in Izotope RX6. Screen capture by the author.

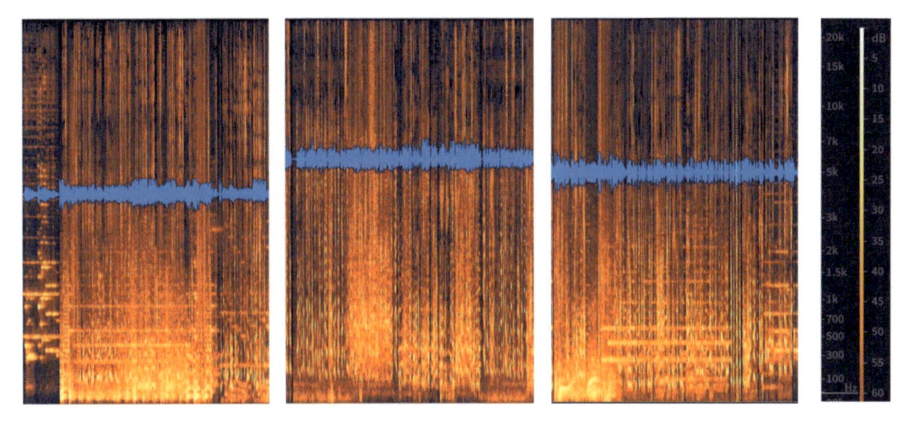

Figure 8. Spectrograms of the openings of *Death in Ice Valley*, *Lost in Larrimah*, and *Someone Knows Something*. Decoded in Izotope RX6. Screen capture by the author.

sorts of unusual elements, including abstract atmospheres and musical details (figs. 7 and 8). In talk-forward podcasts, most of the sonic information naturally falls in the range of human voice from around 125 hertz to 5 or 6 kilohertz (including harmonics), and a spectrogram mostly reveals voices following one another. But with the dramas of recessive epistemology, there are mottled sound compositions dappling the spectrogram with clear notes (see the beginning section of the image of *Death in*

Ice Valley on the left of fig. 8), passages with broadband sound indicating thick barroom soundscapes (there is one about one-third of the way through the spectrogram of *Lost in Larrimah* in fig. 8), passages with low, loud car sounds preserved for effect, and long horizontal bars indicating drone effects across scenes and narrations (across the bottom of the image of *Someone Knows Something* in fig. 8). Backgrounds overwhelm, but they can also take on rudimentary literary functions. Early in *Someone Knows Something*, for instance, Ridgen speculates on birds in the drawings of young Adrien. Then, in the remainder of the series, every time we return to Holmes Lake, the soundscape of the piece is pestered by crows and ravens.

7. The Ludic Chain. True crime podcasts don't die; they run out of episodes. By the end of the Isdal case, we have a number of speculations: a counselor thinks she was a Holocaust refugee who may have worked for Mossad; a former KGB officer thinks she may have been a courier for a spy ring in the Norwegian government; a former spy catcher thinks she died of misadventure, dropping flammable hairspray into a campfire. A final episode lets fans air their speculations, democratizing access to factuality in a way that shows how much this paradigm has drifted from the authority parables of early docucrime. *Someone Knows Something* takes a different tack, focusing on the grief of the family, like narratives of familiar knowing discussed above. *Lost in Larrimah* finds a kind of poetic wisdom in its nonconclusion, ending its speculative ludic chain with a quiet chance scene of a group of wild donkeys in the dark, sounds that reflect on the vanishing not of Moriarty but of the way of life that his town represents. It's an ending at the level of affect, a tangible loss that grazes the ears, a moment of reflection that feels more like a dwelling than a release. The story isn't over, only the game we've been playing with it.

How to capture the collective effect of these features? In thinking through this issue, I would like to rely on a perhaps unexpected text here: *Mimesis*, Erich Auerbach's well-known book on literary representation. Readers of that classic tome will recall that in it the author sets up a binary between two ancient forms of mimetic realism. As exemplars, he selects first a passage from the *Odyssey*, in which Odysseus returns from his long voyage and is recognized by his childhood nurse Euryclea thanks to his scar, and then another from the Old Testament, the Akedah, the story of Abraham's binding of Isaac from Genesis 22. In the first story, narration is offered in a transparent and straightforward fashion, as Auerbach explains, describing all narrative information as "scrupulously externalized," "narrated in leisurely fashion," "orderly, perfectly well-articulated," "clearly outlined," and "brightly and uniformly illuminated" in "a reality where everything is visible," including the feelings and thoughts of characters.[61] Homer's art bears clear meanings, leaving no feeling of inner turmoil or gap. Contrast this style with that of the Bible, whose style sets only certain parts into high relief, leaving others obscure, featuring the "suggesting influence of the repressed" and a "multiplicity of meanings," as well as a preoccupation with the problematic.[62] Unlike Homeric poetry, biblical writing is "fraught with background," which is why it prompts hermeneutic responses among readers trying to fill in what feels like gaps of information and meaning, hence the millennia of exegetical readings of the Akedah that try to explain things the text does not clearly provide. Biblical realism is how writing comes to terms with a "reality" that possesses a quality Auerbach called in German *deutungsbedürftig*—a sense of standing "in need of interpretation."

The podcasts of the epistemological turn, and their obsession with structures of knowing, evoke a world characterized by this need and thus represent a change in radio's realist regime from Homeric narration—that's what is happening in *This American Life*—to a biblical-style version characterized strongly by *deutungsbedürftig*. Podcasts of recessive epistemology are an extreme version of this phenomenon. Importantly, this isn't a retreat from commitment to realism but a transformation in the idea of the world to which the realistic drive is beholden. And it is in some ways a relief to have a kind of radio that expects to be interpreted and critiqued, not one polished and sealed off from exterior thought. It is even possible to think of this as a return to the roots of radiophonic experience. Note how when Auerbach speaks of the Homeric style, his language is full of metaphors of visibility, and there is indeed a sense in which radio forms obeying that realism aspire to the clarity of visual spectacle. But much 2010s podcast storytelling comes from a different tradition, conjuring not the world of radio's positivist champions like John Reith, Bill Siemering, and even Jay Allison, but that of its occultists and mystics, of Upton Sinclair, Aimee Semple MacPherson, Sir Oliver Lodge, Velimir Khlebnikov, and Antonin Artaud, what radio artist Gregory Whitehead calls the dreamland/ghostland of early radio in which the medium was understood to link unknown bodies to one another and to life and death.[63] In that context, podcasting represents radio's return to a world much like the one it knew when the medium emerged a century ago, in the dark.

What We No-No

At the outset of this chapter, I noted how frequently podcasts of this period used the phrase "what we know" when they reached their concluding moments, even if (especially if) these podcasts ended up not knowing much more for certain than when they set out. This predilection signals a prominence to ways of knowing that make these podcasts quite different from many of the American-style radio shows and podcasts that both preceded and surrounded them, the kind in which storytellers (at least at the moment of enunciation) always seem to know what they want to know and how to explain it all. In the earlier, more dominant model, no matter the twists and turns we take in the temporal world of the tape and in the memories of the host about how they learned what they learned, we know that all will be revealed in the foreground of these orderly visible worlds; the expectant ear, patiently awaiting a point, punchline, or empathic catharsis, is seldom cheated. Producers felt they could rely on simple formulas and improvised strategies for cutting down material or generating interest whenever their reporting became inchoate or went in too many directions at once, reaching a point of frustration in the production process that *Radiolab* head Jad Abumrad has memorably called being "lost in the German Forest."[64] Some of these formulas have been spelled out above (more are spelled out in Jessica Abel's *Out on the Wire*), but the details are less crucial than the fact that any story formula was able to accomplish two things: first, training listeners to expect a way out of conundrums; second, satisfying that expectation most of the time, thereby naturalizing this process as the signature deliverable of high-quality audio stories. In these efficiently unloaded narratives, we could feel confident as listeners walking into the forest with our hosts. We may not have a map, but we know they do (even if they don't share it right away), and in finding a way out together we will have acquired a fragment of knowledge of the world around us, braided within a satisfying emotional release and a sense of empathetic connection.

It is upon this certainty that the podcasts of the epistemological turn cast a jaundiced eye, as many of the podcasts discussed above tried to find ways to think outside of it, while never fully offering an alternative paradigm entirely divested of key elements of the American style. That inability to "revolutionize" should not eliminate these works from podcast history. The mood behind them was one of the richest features of narratives of the late 2010s, expressing a renewed interest in contingency, inequities of perception and experience, structural impediments to the flow of "knowing" along lines of race and gender, and aesthetic features that could even suggest morbid transfixion with the impossibility of resolution and containment that enweaved with and animated fears of climate change and other social catastrophes in the period. Some podcasts of the epistemological turn brought a refreshing politicization to audio style; others brought a frank, naturalistic nihilism. Both were things that early models of narrative audio lacked.

And the concerns of the epistemological turn have not vanished. A prominent recent example of a podcast obsessed with structures of knowing is the *New York Times*'s *Trojan Horse Affair*, which encodes in its main text the many years and dead ends of its own investigation, voluptuously explores voluminous archival material, includes direct and repeated dramatization of process between the two journalists Brian Reed and Hamza Syed, and makes extensive reflections on the politics behind journalistic pretensions of remaining detached, especially in societies filled with racist logics. Meanwhile, early on in the pandemic, Pushkin Industries began to release Jill Lepore's remarkable podcast *The Last Archive*, which drew on a rich variety of extant audio archives to examine "how we know what we know, and why it seems, these days, as if we don't know anything at all." Episodes zeroed in on the historical replacement of mystery with fact in court trials, the rise of feminist consciousness raising and the idea of speaking one's truth, the role of credulity and incredulity in P. T. Barnum's hokum, and the history of predicting weather. A kind of *Twilight Zone* for historical epistemology, Lepore's podcast makes use of reenactment, transcript, and even archival sound effects, in this way formally reflecting upon the historical constructedness of knowledge that it nominally explores.

During the epistemological turn and after, narratives "knew" differently because the worlds they mirrored were differently characterized, often as more complex, inequitable, politically calcified, and unyielding to analytical thought; podcasters saw a world more like the one we really live in, and they told tales about it relentlessly. Indeed, there is perhaps no other period in the history of American audio in which more quality soundwork was created directly *about* the deep moral and cognitive frustrations of creating soundwork. Journalists and producers were not waiting until they got out of the German forest to tell their stories; they were broadcasting from inside, and it is no surprise that sometimes they found themselves anxiously contemplating the possibility that forest is all there is.

To conclude this chapter on structures of knowing in narrative podcasts of the late 2010s, I'd like to reflect here on how to think about another variety of structures of knowing: those that are involved in analyzing podcasts, as I have been doing here. For instance, one thing that stands "in requirement of interpretation" in this book has been my use of visualizations in coming to know podcast aesthetics. These are, of course, different ways of knowing what is happening in these works. In chapter 1, I used visual renderings of pitch contours, as measured by bespoke digital audio measurement tools, to explore the vocal crack of Adnan Syed in the first season of *Serial*, a visual-analytical approach that to my knowledge no critic has employed with this well-studied podcast. I also used another digital tool in the subsection just above, spectrograms of recessive epistemology episodes, to provide an understanding of their sound in contrast to other types of podcasts. To create these spectrograms, I used an audio repair program developed chiefly for dialogue editing called Izotope Rx, which employs a fast Fourier transform (FFT) algorithm, a common method for transforming a complex time-domain audio signal into a frequency-domain heat-map-style visualization showing how the track's playback code instructs for the distribution of sonic energy at different frequencies as measured in hertz. You can see at what frequencies the intensities of signal energy are concentrated at any given moment along the left-to-right time axis, or, to put it inversely, you can see how any given frequency's intensity varies across time in a piece of audio across the top-to-bottom frequency axis. FFT-derived images help an editor see anomalies that had entered a mix either by accident or by design and help them to identify elements they may want to change: an annoying bird call caught by the mic during exterior dialogue capture that you can see and remove with the click of a mouse; rumbles typical of distant cars to minimize using EQ or a gate in the processing; a visual rendering of how musical elements interact with vocal frequencies and loudness over the duration of a mix in case that needs adjustment; or how the introduction of reverb to one element might affect others at particular moments. Any FFT-based heat-map-style audio program can do this work, turning a signal scape into a landscape you can repair, trim, prune, and shape.

On one level, it is not all that odd to approach the study of podcasts through a visual interface. Indeed, ten years ago, Michele Hilmes had already realized that podcasting introduced screen interactivity to the world of radio for the first time, and this phenomenon has spawned a number of critical works that look at podcasting through its online screen-based communities by sifting through websites of supplements and chatter.[65] But these analyses tend to focus on how listeners interact with screens to access content, not how producers interact with screens to generate content. The difference between the screen as a host of a website or interactive fan forum and the screen as a locus for an expensive, sophisticated program like Izotope Rx is stark. The latter seems to transform the piece as a work of sound into a map, one that is not easy to read and has been largely "for" engineers and sound designers rather than fans or critics; although FFT algorithms are used to decode audio signals in many situations, to my knowledge, no publicly available podcatcher app displays podcast audio in this way for ordinary listeners. Spectrographic criticism rooted in such images naturally represents a new way of knowing a piece, showing, for example, the relative prevalence of voice over music at certain moments, specifying the recurrence of motifs, the ratio of studio recording to field recording, the way elements work over a duration or cross scenes, and so forth. I reflect on this here because it is obvious that to employ visual remediation of a piece of sonic media in a critical context is to superimpose an alternative interpretive structure upon the audio, a way of knowing, and my adoption of this method to analyze works therefore suggests my own methodological participation in the very epistemological turn that this chapter has endeavored to describe.

For some insight on the meaning of this move, and on what podcast studies tell us about structures of knowing more broadly, I would like to back up and approach the phenomenon of podcast visualization from the other end, looking at how these pieces are mixed on screens, a practice that has its own hidden politics that were simmering during the period. In the winter of 2021, *The Verge* published an article (derived from a Bello Collective piece), circulated widely among audio producers on Twitter, that centered on what outsiders might not realize is a reason why organizations in the audio industry had failed to reach the level of diversity, equity, and inclusion they often committed to in florid mission statements.[66] The article speculated that a secret culprit was the industry's reliance on Pro Tools, a powerful but complex and expensive digital audio workstation (DAW) that had long been standard in film and television work and that was increasingly a job requirement in audio. Like other DAWs, Pro Tools generates a visual interface on which producers, designers, and engineers mix music, sound effects, and studio and field recording as discrete pieces of coded audio using sophisticated automations (allowing you to vary any parameter—volume, reverb, stereo position, etc.) and a vast library of plug-ins. To many audio people it felt like making anything possible when it comes to making stories. Like film editors a generation before them, radio editors discovered many advantages that came with DAW work, including nondestructive editing opportunities and random access to any position in the track, letting editors find files and play them without rewinding or fast-forwarding, just a few of the features that media scholar Katherine Quanz has identified about these systems.[67] But in the audio world, many aspects of these workstations could also be unnecessarily labyrinthine to producers who generally did not require many of its affordances. *The Verge*'s author went after this discrepancy, calling Pro Tools "monstrous" and "nightmarish."

There are many DAWs like Pro Tools, and their rise in the twenty-first century affected both radio and podcasting alike in a way that was generational rather than medium specific. DAW-based audio production likely has deeper implications for the way audio storytelling is conceived and materially constructed, from the producer's perspective, than even the very rise of podcast distribution itself. You could liken the advent of DAWs in the 2010s to the invention of devices like the Dramatic Control Panel, the first mixing board used for radio drama, which, as Jeremy Lakoff has argued, accelerated the divestment of audio drama from stage drama in the late 1920s by allowing for the possibility of multiple channels and studios to be used for a single play, making the station into a kind of instrument.[68] And just as the panel was like learning a new instrument, so did the advent of DAW-based editing change the way the trade was conceived and trained for. For many prominent creators in the audio field of the 1990s and early 2010s, it was no distant memory to recall the use of material magnetic tape manually cut and spliced to form a story for air. For some audio practitioners in the early 2000s and 2010s, versions of simple DAWs such as Hindenburg and Audacity were easy to master and could be found at inexpensive prices online, while others could get a head start in the industry just relying on audio components of larger standard software suites used in colleges or in business like Apple's GarageBand and Adobe's Audition. By the 2020s, this was no longer enough, as job postings for those aspiring to work in the area often listed a Pro Tools expertise requirement, and even seasoned producers were forced to retrain. What had long been touted as a career that anyone could get into with a handheld Zoom H4 recorder, a laptop, and cheap software was increasingly a field whose main gatekeeper

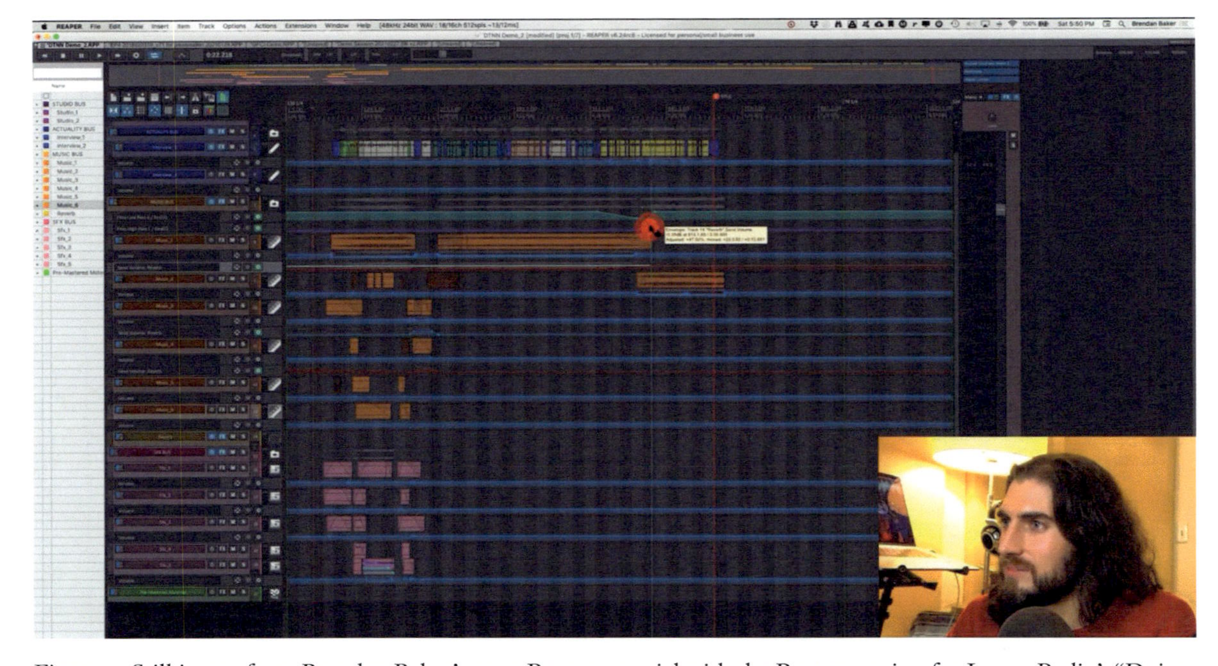

Figure 9. Still image from Brendan Baker's 2021 Reaper tutorial with the Reaper session for Love+ Radio's "Doing the No-No" session in background. Screen capture by the author.

was an unwieldy, non-purpose-built digital workspace, and it took thousands of dollars in an online course, or hundreds of hours of job training, to attain facility with the platform. That situation presented barriers to creators from backgrounds historically excluded from expensive technologies and unpaid internships, including women, people of color, and others marginalized due to class, sexual expression, ability, or language. For talk-based programs, this was less of an acute problem, with the emergence of programs such as Descript, a word processor for audio that allows you to edit audio as text and vice versa, which was marketed at podcast conferences in this period. But for story-based narrative pieces that sounded like the highly designed podcasts that gained prominence in the 2010s, deeper sophistication meant higher barriers.

Did it have to be this way? Partly in response to the article in *The Verge*, in April 2021 the widely admired podcast producer, sound designer, and director Brendan Baker hosted a series of free online tutorials for an alternative DAW called Reaper, which, in addition to being relatively affordable, brought with it all the capabilities and visualizations a producer could ask for when it came to editing dialogue, adding sound textures and music, and managing loudness. Baker also distributed a system of presets that optimized the software for the kind of audio he produced, trying to explain his own workflow in the hope it could be adapted by others who lacked the funds for Pro Tools and the training courses it took. With Reaper configured correctly and with a few other inexpensive plug-ins that Baker also provided, you could edit, design, mix, and master with the same sophistication as Pro Tools at a fraction of the cost. Baker was not the first to call attention to this alternative DAW. Six years earlier, *Here Be Monsters* producer Jeff Emtman had published a similar set of free tutorials online for an earlier version of the same program. Something of an evangelist for the software, Emtman pushed Reaper's possibilities to the limits, and he mentored and inspired many other key producers early in their careers.

Baker's revelatory three-hour video session on Reaper for podcasting—originally broadcast live via Zoom, with an edited version still available on YouTube at the time of writing—concludes with a demonstration of the last piece he made while he was a producer at *Love + Radio*, which, by coincidence, is the "Doing the No-No" story that I discussed in the section above on formal structures of knowing, the story that starts out as a manifesto of sorts by rogue bio artist Adam Zaretsky, keen to explain his many thoughts on transgenics as an art form, but that over time becomes

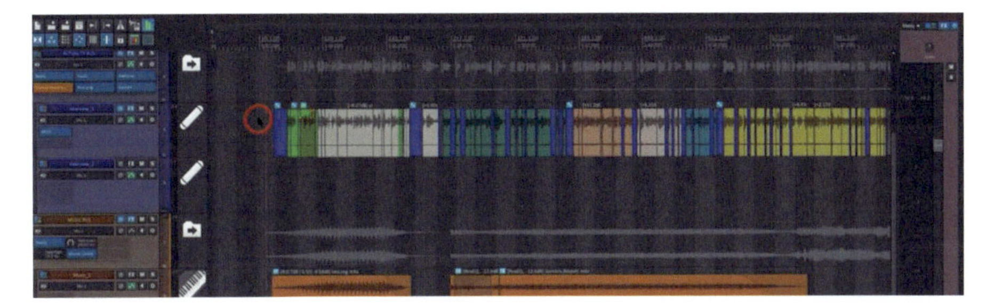

Figure 10. Still image from Brendan Baker's Reaper session, dialogue track detail. The colors and segmentation indicate edits and crossfades being formed using many separate raw interviews compiled to create the final track. Screen capture by the author.

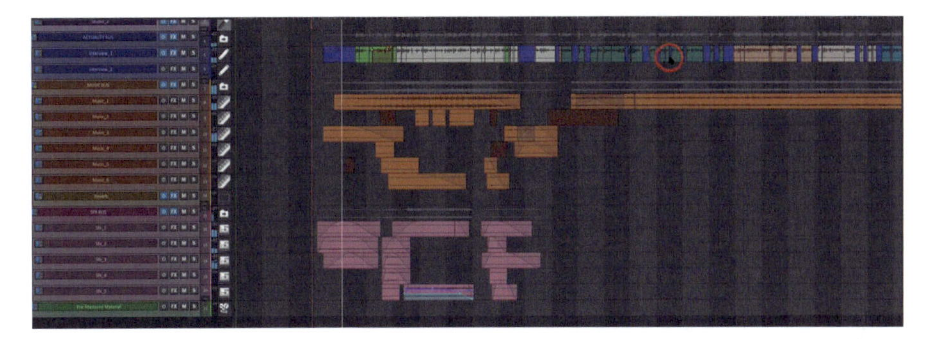

Figure 11. Still image from Brendan Baker's Reaper session, showing relationships between dialogue items (top section), music (middle tracks in orange) and sound effects (lower section in pink).

Britt Wray's critique of the biopower that Zaretsky seemed eager to wield on others.[69] The podcast shifts from structuring knowledge about art to structuring our knowledge about its practitioner. It is fascinating that Baker's Reaper tutorial is not unlike a Zaretsky piece, in that both showcase a highly technical editing process (genes for Zaretsky, audio for Baker) that had for the most part largely been used to "perfect" something in normative terms (bodies, stories), and both were trying to turn the technology to outsiders (queer people, producers of color) in a Promethean way, in the hope of empowering more inclusive art production (bio sculpture, podcasting). The analogy ends there, of course, as Baker's generous sessions reverse the self-interestedness of Zaretsky's practice; you could even say that Baker's project to open up audio vindicated the faith in making art that respects the "radical porousness of reality" that Zaretsky professed but failed to practice.

I also think the exercise of watching this particular piece on screen, after a long tutorial about how the session was set up, can help illuminate the crucial importance of structures of knowing and how to approach them. Watching the video, having already heard the piece, it is as if one structure of knowing intervenes on another, like a subduction between tectonic plates breaking the landscape, revealing the structure of both layers. My argument is that—in podcast studies and anywhere else—a critical approach to a structure of knowing is helped by the fact that they are always multiple and stacked, with one vying against another for attention. Look at the single track of Zaretsky's voice (fig. 10) on the screen in Baker's video as it plays, for example, and you can see dozens of cuts and splices that represent in-DAW edits of the raw tape so numerous that it looks almost like genetic code. The auditory experience is revealed to be thoroughly prestructured by a meticulous editing process often inaudible to the ear, an experience that lends a doubleness to the listening experience. It is a little like watching a film and noticing for the first time that there are cuts between shots, that there have *always* been cuts between shots.

Then look at the larger structure of the session as a whole (fig. 11) and you see a total of five color codes for track types of the left (purple for studio host dialogue, blue for field interviews, orange for music, pink for sound effects, green for other materials), with all the audio items corresponding to these categories in each track placed on the right on the timeline. In the video, as the play head marker moves across the screen, Zaretsky narrates a gruesome metaphor from Greek mythology about the birth of Aphrodite from the foaming seas around the castrated penis of Uranus. While this narration is taking place, a whole set of musical tracks intervene, rise, and recede, while a section of sound effects elements also operate like a miniature abstract radio drama, illustrating Zaretsky's story with splashes and cavernous reverb. What sounded at first to be one narrative looks on the screen like three stacked narratives that interact. If the cuts in the interview track suddenly look like slashes in the narrative—a hundred little dismemberments of its own—then the wordless sound drama is a palliative cartouche surrounding the verbal one, returning it to a sense of harmony, one structure resting on the other.

Viewing these elements interact in the landscape of a DAW session really transforms the sense of the tightness of tolerances in the mix, but it also decomposes the mix before our eyes. Watching Baker's video feels magical to a layperson, like watching mixed paint returning to its component colors. Theorists of sound have long been wont to focus on how, unlike images, sounds are irreducible, belonging to a "philosophy of mixed bodies" exceeding frames, walls, rooms, and containers that readily contain our visual experiences; this is a subsidiary of the notion that sound is bounded by time rather than space, a tenet of the "audiovisual litany" that Jonathan Sterne once identified as a way in which discourses of sound have historically over-differentiated these two senses and tacitly defended a paradigm of transcendence over materiality.[70] The whole idea of a DAW is to make sound operate by the same principles as sight—or to reveal how it operated that way all along—by pouring sound into visual equivalents of containers. To ignore that and to insist that only the final amalgam represents a "true" version of the piece would be a shame, wasting a glimpse into the tracking architecture and mixing process for podcasts in this period, facts we can only speculate about when the audio is in our ears. Viewing them, you get a sense you are learning a secret structure of the story, watching a landscape being painted, experiencing not an artificial or degraded "representation" of an original work of audio but the visual representation that is prior to that "original." For these reasons, I have long believed it would be worthwhile to preserve DAW sessions in archives, since in many cases it is in these programs where dialogue and design decisions are made, and these files represent a trace of that generative process. (For an interactive version of such a mini-DAW, see https://doi.org/10.3998/mpub.11751593.cmp.13)

To summarize, the theory of structures of knowing, when pursued in this way, leads almost inevitably to an acute sense of the heterogeneous multiplicity of the digital object itself and to our inability to settle on only one version of it as an entity capable of hosting a unitary, thorough reading. What we know about "Doing the No-No" becomes a rather complex matter. The narrative contains two vying structures of knowing between Zaretsky and interviewer Britt Wray when it comes to bio art as a subject matter. Its DAW session contains several structures as well that both mirror the audio version and also each provide a different sense of a material manipulation that we hadn't been entirely aware of before. The session as a whole can also be viewed as a structure of knowing unto itself, with various tracks structuring others, and its playback yet another. The same is true for spectrograms, pitch trackers, or any visualization device that reveals structures inside what we hear precisely by being something we cannot hear but only see. What's more, as I have detailed above, there are inequities of access at all stages—between marginalized podcasters and DAW wizards, between braggadocious artists and their coerced collaborators, between creators and the scholars who try to get into their work—which suggests that structures of knowing are hardly bloodless, abstract notions but are instead shaped by political and material conditions of hierarchy, order, and social difference. As Michel Foucault often reminded us in many works on the relationship between power and knowledge, power is not just a phenomenon of the hand of authority but a system distributed throughout the social body, one that works best not when it prevents behaviors and reactions but when it creatively forms them.[71]

This has been a long conclusion; let me reshape it into a narrower form. I am proposing an argument that has three stages. The first is that one vital way to understand the sound of a podcast is, ironically, to look directly at it, both through the interfaces that producers themselves use to look at it on-screen in DAWs prior to release and as we ourselves might look at it through digital tools like pitch trackers, FFT-based spectrograms, and other methods. The second stage builds on the first, but it also undermines it at the same time: to look at a podcast is to know that there is no singular "podcast" as a unitary object for us to study, and it would be a mistake to follow the reflex to see these visualizations as reified versions of a true object, lurking in underlying code. On the contrary, a particular episode "is" the intersection of several perfectly legitimate and worthy ways of knowing that episode—as auditory experience, as an image, as data, as experiences that differ depending on the listener—each of which prompts a different way of accounting for the text's basic existence and poetics. Finally, in light of these first two points, I want to propose that the use of digital tools for podcast studies in this chapter may be properly considered to be a late-breaking phenomenon of the epistemological turn in podcasting, a form that belongs to a juncture in podcast history (and podcast studies) characterized by a heightened awareness about who gets to know what and how, about stages of production we may never see, lost in scattered, atrophying hard drives. Digital affordances reveal the very object we study to be irreducibly multiple, hierarchized and layered, possibly endangered, and at any rate in obvious epistemological tension with itself. They show us what is both rich and daunting when it comes to the task of unpacking what any podcast knows, how it knows—and how we know what it knows. Far from finding a way out, the task of criticism is to find a way into the German forest, where an extended stay might not be so bad after all.

Audio Works Discussed in Order of Appearance in the Text—Born Podcast/Hybrid

The Heart (Kaitlin Prest, Mermaid Palace, Radiotopia, February 5, 2013–February 23, 2022, https://www.theheartradio.org/all-episodes).

The Allusionist (Helen Zaltzman, January 14, 2015–present, https://www.theallusionist.org).

West Cork (Sam Bungey, Jennifer Forde, Audible Originals, February 8, 2018, https://feeds.acast.com/public/shows/620c6279-d89b-4719–947d-a5f4b47b44eb).

Forest 404 (Timothy X Atack, Becky Ripley, BBC Radio 4, April 4, 2019–May 29, 2019, https://www.bbc.co.uk/programmes/p06tqsg3/episodes/guide).

Sonic Voyages (Kara Oehler, *New York Times Magazine*, September 21, 2018, https://www.nytimes.com/interactive/2018/09/21/magazine/voyages-travel-sounds-from-the-world.html).

The Stoop (Leila Day, Hana Baba, Radiotopia, July 18, 2017–present, http://www.thestoop.org/).

Radio Atlas (Eleanor McDowall, Radio Atlas, July 9, 2017–October 23, 2022, https://www.radioatlas.org/feed/podcast/).

Constellations (Michelle Macklem, Jess Shane, Constellations, July 25, 2017–present, https://podcasts.apple.com/us/podcast/constellations/id1263430011).

This American Life (Ira Glass, WBEZ Chicago, November 17, 1995–present, https://www.thisamericanlife.org/archive).

Home of the Brave (Scott Carrier, December 17, 2015–March 21, 2022, https://homebrave.com/).

Serial (Sarah Koenig, Julie Snyder, et al., WBEZ Chicago, October 3, 2014–present, https://serialpodcast.org/).

Up and Vanished (Payne Lyndsey, Tenderfoot TV, August 7, 2016–May 26, 2022, https://podcasts.apple.com/us/podcast/up-and-vanished/id1140596919).

Limetown (Zack Akers, Skip Bronkie, Two-Up Productions, July 29, 2015–December 3, 2018, https://twoupproductions.com/limetown/podcast).

Death in Ice Valley (Marit Higraff, Neil McCarthy, Norwegian Broadcasting Corporation, BBC World Service, April 15, 2018–June 6, 2021, https://www.bbc.co.uk/programmes/p060ms2h/episodes/downloads).

Someone Knows Something (David Ridgen, CBC Podcasts, March 1, 2016–July 13, 2022, https://www.cbc.ca/listen/cbc-podcasts/128-someone-knows-something).

Nocturne (Vanessa Lowe, October 20, 2014–present, https://nocturnepodcast.org/nocturne-episodes/).

Love + Radio (Nick van der Kolk, October 18, 2005–present, https://www.wbez.org/shows/loveradio/700796df-7723–4ace-9c80–3121ca615605).

Benjamen Walker's Theory of Everything (Benjamen Walker, Radiotopia, April 19, 2013–present, https://theoryofeverythingpodcast.com/episodes/).

UnCivil (Jack Hitt, Chenjerai Kumanyika, Gimlet Media, September 20, 2017–October 27, 2020, https://gimletmedia.com/shows/uncivil).

Scene on Radio (John Biewen, Center for Documentary Studies at Duke University, September 1, 2015–August 10, 2022, https://www.sceneonradio.org/episodes/).
BackStory (Ed Ayers, Brian Balogh, Nathan Connolly, Joanne Freeman, Peter Onuf, Virginia Humanities, September 4, 2015–July 3, 2020, https://www.backstoryradio.org/episodes).
American History Tellers (Lindsay Graham, Wondery, January 3, 2018–present, https://wondery.com/shows/american-history-tellers/).
Slow Burn (Leon Neyfakh, Andrew Parsons, Slate Podcasts, November 28, 2017–present, https://slate.com/podcasts/slow-burn/s7/roe-v-wade).
The Frost Tapes (Wilfred Frost, David Paradine Productions Ltd., October 5, 2020–July 31, 2022, https://podcasts.apple.com/us/podcast/the-frost-tapes/id1532912344).
Bag Man (Rachel Maddow, MSNBC, October 29, 2018–December 3, 2018, https://www.msnbc.com/bagman).
Fiasco: Bush v Gore (Leon Neyfakh, Luminary, May 22, 2019–November 6, 2019, https://luminarypodcasts.com/listen/leon-neyfakh/fiasco-luminary-premium/8607a31a-cab7-4c9f-9df8-b6110ba93b52?country=US).
StoryCorps, "Remembering Stonewall: 50 Years Later" (Dave Isay, NPR, June 25, 2019, https://storycorps.org/podcast/remembering-stonewall-50-years-later/).
Making Gay History (Eric Marcus, Making Gay History, October 13, 2016–present, https://makinggayhistory.com/).
Louder Than a Riot (Rodney Carmichael and Sidney Madden, NPR, Sept 16, 2020–May 20, 2022, https://www.npr.org/podcasts/510357/louder-than-a-riot).
Shots in the Back: Exhuming the 1970 Augusta Riot (Sea Stachura, George Public Broadcasting, June 28, 2020–October 19, 2020, https://www.npr.org/podcasts/824570049/shots-in-the-back-exhuming-the-1970-augusta-riot).
The Memory Palace (Nate DiMeo, Radiotopia, November 12, 2008–present, https://thememorypalace.us/episodes/).
The Kitchen Sisters Present Fugitive Waves (Davia Nelson, Nikki Silva, Radiotopia, February 5, 2014–April 27, 2021, https://exchange.prx.org/series/33538-fugitive-waves?order=newest_first).
The Kitchen Sisters Present, "The Keepers" (series) (Davia Nelson, Nikki Silva, Radiotopia, September 5, 2018–March 24, 2020, https://kitchensisters.org/keepers/).
Floodlines (Vann R. Newkirk II, *The Atlantic*, March 11, 2020, https://www.theatlantic.com/podcasts/floodlines/).
9/12 (Dan Taberski, Pineapple Street Studios, Wondery, September 8, 2021–October 13, 2021, https://podcasts.apple.com/us/podcast/9–12/id1581684171).
Cocaine and Rhinestones (Taylor Coe, October 24, 2017–February 15, 2022, https://www.cocaineandrhinestones.com/).
You Must Remember This (Karina Longworth, April 16, 2014–present, http://www.youmustrememberthispodcast.com/episodes).
Dan Carlin's Hardcore History (Dan Carlin, October 28, 2015–present, https://www.dancarlin.com/hardcore-history-series/).
Phoebe's Fall (Richard Baker, The Age, *Sydney Morning Herald*, September 21,

2016–December 15, 2016, https://podcasts.apple.com/us/podcast/phoebes-fall/id1155393027).

Hanging: The Boy in the Barn (Julia Prodis Sulek, *Mercury News*, August 24, 2017–September 3, 2017, https://extras.mercurynews.com/hanging/podcast.html).

The Ballad of Billy Balls (iO Tillett Wright, iHeartPodcasts, March 28, 2019–July 11, 2019, https://www.theballadofbillyballs.com/).

The Missing Cryptoqueen (Jamie Bartlett, BBC Radio 5 Live, September 19, 2019–December 22, 2022, https://podcasts.apple.com/us/podcast/the-missing-cryptoqueen/id1480370173).

Three Rivers, Two Mysteries (Michael A. Fuoco, Ashley Murray, *Pittsburgh Post-Gazette*, October 24, 2017–December 26, 2017, https://newsinteractive.post-gazette.com/threeriverstwomysteries/).

The Trojan Horse Affair (Brian Reed, Hamza Syed, Serial Productions, New York Times Company, February 3, 2022, https://www.nytimes.com/interactive/2022/podcasts/trojan-horse-affair.html).

Unsolved (Gina Barton, *Milwaukee Journal-Sentinel*, November 3, 2015–April 24, 2019, https://projects.jsonline.com/topics/unsolved/season-three/the-devil-you-know.html).

Unresolved (Michael Whelan, Unresolved Productions, October 7, 2015–present, https://unresolved.me/stories).

Unfinished (Taylor Hom, Ash Sanders, Neil Shea, Sarah Ventre, Witness Docs, June 28, 2020–August 23, 2022, https://www.witnesspodcasts.com/shows/unfinished-deep-south).

Invisibilia (Lulu Miller, Alix Spiegel, NPR, January 9, 2015–present, https://www.npr.org/programs/invisibilia/).

Here Be Monsters (Jeff Emtman, KCRW, January 1, 2012–present, https://www.hbmpodcast.com/episodes).

In the Dark (Madeleine Baran, APM Reports, September 7, 2016–June 11, 2020, https://features.apmreports.org/in-the-dark/).

Caliphate (Rukmini Callimachi, New York Times Company, April 19, 2018–December 18, 2020, https://caliphate.simplecast.com/episodes).

Criminal (Phoebe Judge, Radiotopia, January 28, 2014–present, https://thisiscriminal.com/).

The Organist (Andrew Leland, KCRW, April 15, 2014–April 18, 2019, https://www.kcrw.com/culture/shows/the-organist).

Have You Heard George's Podcast? (George the Poet, BBC Radio, August 31, 2019–September 15, 2021, https://www.bbc.co.uk/programmes/p07915kd/episodes/downloads).

Everything Is Alive (Ian Chillag, Radiotopia, July 16, 2018–present, https://www.everythingisalive.com/).

The Organist, "The Narrative Line" (Andrew Leland, KCRW, April 4, 2019, https://www.kcrw.com/culture/shows/the-organist/the-narrative-line).

Invisibilia, "The End of Empathy" (Alix Spiegel, NPR, April 12, 2019, https://www.npr.org/programs/invisibilia/712280114/the-end-of-empathy).

Radio Ambulante (Daniel Alarcón, NPR, November 22, 2016–present, https://radioambulante.org/en/episodes).

Last Seen (Kelly Horan, Jack Rodolico, WBUR, September 14, 2018–present, https://www.wbur.org/podcasts/lastseen/archive).

Somebody (Shapearl Wells, Topic Studios, The Intercept, The Invisible Institute, iHeartRadio, March 31, 2020–June 12, 2020, https://theintercept.com/podcasts/somebody/).

The Clearing (April Belascio, Pineapple Street Studios, Gimlet Media, July 16, 2019–August 29, 2019, https://gimletmedia.com/shows/the-clearing/episodes#show-tab-picker).

Missing & Murdered: Finding Cleo (Connie Walker, CBC Podcasts, October 25, 2016–April 2, 2018, https://www.cbc.ca/radio/findingcleo/click-here-to-listen-to-missing-murdered-finding-cleo-1.4557887).

Radio Diaries (Joe Richman, Radiotopia, May 16, 2013–present, https://podcasts.apple.com/us/podcast/radio-diaries/id207505466).

This American Life, "Abdi and the Golden Ticket" (Leo Hornak, WBEZ Chicago, July 3, 2015, https://www.thisamericanlife.org/560/abdi-and-the-golden-ticket).

Latino USA, "The Return" (Sayre Quevedo, Futuro Media, PRX, December 14, 2018, https://www.latinousa.org/2018/12/14/thereturn/).

Ear Hustle (Nigel Poor, Earlonne Woods, Rahsaan "New York" Thomas, Radiotopia, June 14, 2017–present, https://www.earhustlesq.com/listen).

No Feeling is Final (Honor Eastly, ABC Radio National, September 17, 2018–October 18, 2018, https://www.abc.net.au/radio/programs/no-feeling-is-final/episodes).

Unfinished: Deep South (Taylor Hom, Neil Shea, Witness Docs, June 28, 2020–August 23, 2020, https://www.witnesspodcasts.com/shows/unfinished-deep-south).

Radiolab (Jad Abumrad, Robert Krulwich, WNYC, May 25, 2012–present, https://radiolab.org/episodes).

Studio 360 (Kurt Andersen, PRI, *Slate*, October 31, 2016–February 27, 2020, https://www.npr.org/podcasts/381444899/pri-studio-360).

Snap Judgment (Glynn Washington, Snap Judgment Studios, PRX, July 2010–present, https://snapjudgment.org/podcast-episodes/).

Radiolab, "Yellow Rain" (Pat Walters, WNYC, September 24, 2012, https://radiolab.org/episodes/239549-yellow-rain).

S-Town (Brian Reed, Serial Productions, March 28, 2017, https://stownpodcast.org/).

Love + Radio, "Doing the No-No" (Brendan Baker, Britt Wray, December 2, 2016, https://loveandradio.org/2016/12/doing-the-no-no/).

Lost in Larrimah (Caroline Graham, Kylie Stevenson, *The Australian*, April 27, 2018–April 15, 2022, https://podcasts.apple.com/us/podcast/lost-in-larrimah/id1377413462).

This American Life, "Right to Remain Silent" (Ira Glass, WBEZ Chicago, September 10, 2010, https://www.thisamericanlife.org/414/right-to-remain-silent).

The Last Archive (Jill Lepore, Sophie McKibben, Ben Naddaff-Hafrey, Julia Barton, Pushkin Industries, May 2020–present, https://www.thelastarchive.com/).

Audio Works Discussed in Order of Appearance in the Text—Born Radio

Ghetto Life 101 (David Isay, LeAlan Jones, Lloyd Newman, Sound Portraits Productions, NPR, May 18, 1993, https://storycorps.org/stories/ghetto-life-101/).
Teenage Diaries, "Growing Up with Tourette's" (Joe Richman, Josh Cutler, 1996, http://talkinghistory.org/richman.html).
Police Headquarters (Bruce Eells Associates, West Coast NBC, 1932, https://archive.org/details/OTRR_Police_Headquarters_Singles).
Gang Busters (Phillips H. Lord, CBS, January 15, 1936–November 27, 1957, https://archive.org/details/gang-busters-1955-04-02-885-the-case-of-the-mistreated-lady).
Calling All Cars (James E. Davis, CBS West Coast, Mutual-Don Lee, November 29, 1933–September 8, 1939, https://archive.org/details/OTRR_Calling_All_Cars_Singles).
The FBI in Peace and War (Betty Mandeville, Max Marcin, CBS, November 25, 1944–September 28, 1958, https://archive.org/details/FBI_In_Peace_And_War).
This Is Your FBI (Jerry Devine, ABC, April 6, 1945–January 30, 1953, https://archive.org/details/OTRR_This_Is_Your_FBI_Singles).
Dragnet (Jack Webb, NBC, June 3, 1949–July 26, 1957, https://archive.org/details/OTRR_Dragnet_Singles).

CHAPTER 3

The Arts of Amnesia

What I Heard at the Revolution

One morning in November 2015, at the Third Coast International Audio Festival's Filmless Festival in Chicago, I walked into a rented high school gymnasium where radio producers and fans were packed on a series of risers in a makeshift black box–style theater to attend "The Revolution Will Not Be Televised: Radio Drama for the 21st Century." Now extinct, the Filmless Festival offered in-depth listening events blending radio with moving images and other experiments for a few hundred attendees from the United States and abroad, at a time when a significant fraction of the emerging creators in the field could still fit in a single bar. Running in off years when the main Third Coast conference was not being held, the Filmless Festival emphasized Third Coast's aspiration to offer audio the same imprimatur as a global film festival gave to movies, to be the "Sundance Festival of Radio" as the organizers often referred to it.[1]

The "Revolution" session was led by Ann Heppermann from New York City and Martin Johnson from Sweden, two award-winning producers who had just launched *Serendipity*, the podcast of the Sarah Awards (so named due to its hosting organization, Sarah Lawrence College), which thrived as a podcast and a promoting organization from the early days of *Serial* to the pandemic, dedicated to cultivated emerging podcast fiction through initiatives and competitions. "It's time for audio fiction to have its own red carpet," as the Sarah Awards website put it.[2] The experimental show rewarded fresh methods and points of view, particularly works that fell into the "audio fiction" realm more than the "radio drama" category, a distinction that suggested an alignment with a loosely defined artistic and literary sensibility, setting it apart from, on the one hand, theater-based radio drama groups like *L.A. Theatre Works* and BBC adaptations from literature and, on the other hand, the world of independent genre-based fiction like the faux-community broadcast supernatural series *Welcome to Night Vale*. At the same time, the "21st Century" part of the session's title emphasized that, very much like *Theatre Works* and *Night Vale*, the program emphatically sought distance from the hokey styles of vaguely recalled audio of the previous century by means of elevation, irony, or new means of execution.

Around that time on the *Serendipity* podcast, you could hear, for example, Chris Brookes's "Bannerman Quartet," based not on a written script but on an installation of four recordings in a public park in Newfoundland, or Brie Williams's "Status," a monologue set in a series of Facebook updates. In Lea Redfern and Rijn Collins's "Almost Flamboyant," a woman strikes up a friendship with a stuffed flamingo with a flaring growl like Tom Waits, over a soundscape like a deconstructed Waits song—a steamy alley in Chinatown, closely miked cigarettes and birds, workmen tossing out furniture from an old bankrupt nightclub in a dingy alley. Of the many budding organizations nurturing new audio drama in the United States in 2015, *Serendipity* had perhaps the best chance to attract curious writers and producers with an experimental sound idea that sat outside of nonfiction storytelling but that overlapped with prestige nonfiction radio and podcasting, at least in terms of its expected audience.

Serendipity also highlighted the work of other emerging groups dedicated to radio fiction, including Jonathan Mitchell's *The Truth*, the single fiction podcast most consistently cited in podcast recommendation lists of the 2010s, one often credited—and not without justice—with reinventing short-form audio fiction.[3] In 2015 it was perhaps most famous for the play "Tape Delay," the first outright audio drama to be featured on *This American Life* back in 2012, the same year as the first episode of *Night Vale*, two "watershed moments" for audio drama in that decade. Mitchell's play surrounds a phone conversation before an evening date between two characters, Ben and Erica, that sounds quite different depending on whose audioposition we share at a given moment. We begin positioned with Ben as he waits for Erica at a bar. She calls to apologize for being late, but his nervous, affable responses put her off. In the end she refuses to meet him and walks away from him on the street. Ben is perplexed, until he replays a tape he made of the phone conversation captured by his phone and hears that his voice sounds boorish and toxic from her end. For our part, we think that we are hearing an accurate re-presentation of the same scene, only from Erica's audioposition, but actually there are two different versions of the colloquy in the audio play. Ben uptalks in the first version, his voice rising from 160 to 250 hertz over the course of many of his key sentences (see fig. 12). Later, when we hear him on tape, the pitch goes the other direction, starting high and then dropping; he also speaks much more slowly.

This *trompe l'oreille* is especially interesting because the main story goes on to explore Ben's fetishistic relation to Erica's voice, as he insidiously reedits tape of his conversation with her into a more gratifying version of their interaction, which is just what producer Jonathan Mitchell has done with Ben's voice for us. Mitchell's plays of this period often explored the idea of the edited voice: in "The Extractor," an entrepreneur finds a way to eavesdrop on any conversation in history based on sonic traces embedded in wood; in "Sylvia's Blood," voices of ritually cursed characters transform into hectoring clones; in "Do the Voice," a cartoon actor's famous character voice returns to her unbidden; in "Sleep Some More," a freshman spouts essay-level prose while fast asleep. By way of these rich, philosophical fantasies of cursed speech, the show married its two methodological roots in musique concrète and improv comedy, investing more deeply than perhaps any other modern audio group in making work that theorized the automated voice as such.[4]

At the Third Coast session, *Serendipity* also brought recognition to the work of Kaitlin Prest's *The Heart*, a show that rethought the relation between fiction and nonfiction and that also won the prestigious Prix Italia that year with "Movies in Your Head," about a half-dreamed erotic relationship with a woman (played by Prest collaborator Mitra Kaboli) whom the protagonist met in the subway. In the piece, Prest's imagination turns into manic music, a sequence designed by composer Shani Aviram around echoing thoughts of desire and second guesses that take the form of imagined text messages and voicemails that grow increasingly frantic. *Serendipity* later gave an award to a similar *Heart* piece, "Strangers in a Small Café," which tracks the thoughts and conversations of millennial coffee shop goers, a kind of *Under Milk Wood*, if Dylan Thomas's famous 1954 radio play had been about neurotic Brooklyn hipsters instead of bawdy Welsh villagers. In Chicago at the Third Coast session, Heppermann and Johnson played part of "Strangers," as well as their own recent collaboration, "Every Heart Has a Limited Amount of Heartbeats," the story of a chance encounter between two strangers in New York that prompts reflections on mortality through whispers, pops, surface noise, actualities, interviews,

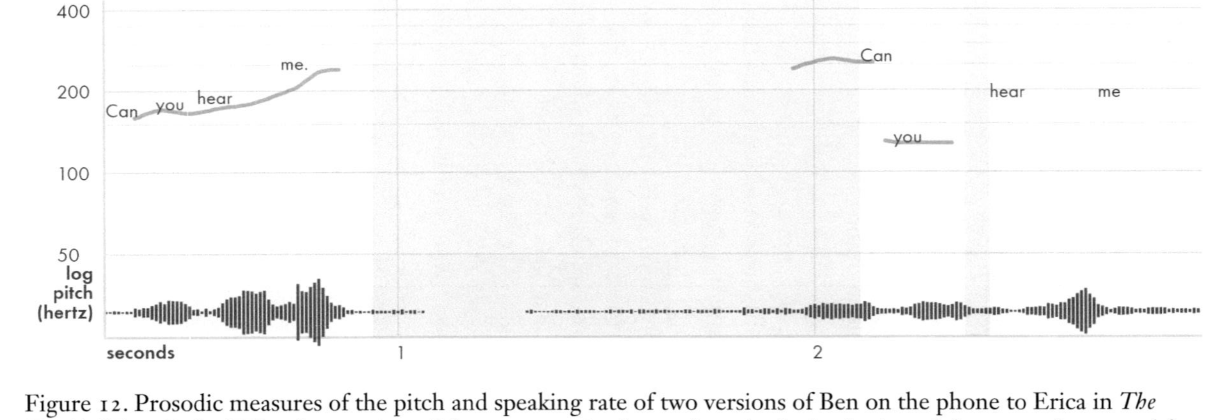

Figure 12. Prosodic measures of the pitch and speaking rate of two versions of Ben on the phone to Erica in *The Truth*'s "Tape Delay." The first version of the utterance is what we hear in the beginning of the scene, the second from the supposedly "objective" tape on his phone of that utterance.

music, and memories, all layered over a double soundscape that culminates in imagined phone calls with the dead. In a 2017 online forum about the potential for podcasting for new radio art, producer Gareth Stack caught the spirit of these pieces in an optimistic comment about "Movies," speaking of an emerging style that merged "hypnagogic sound design with fragments of real and fictional memory" and "fling[ing] time up into the air like pizza dough," thereby turning the problematic isolation and linearity of podcast listening into an advantage.[5]

This was the first time, to my knowledge, that such a challenge for audio fiction came to an event like Third Coast. From its founding fifteen years earlier by Johanna Zorn and Julie Shapiro, Third Coast had been a Mecca for producers from the United States, often likened to a family meeting or group hug for public radio. By the 2010s, a typical meeting had upwards of eight hundred attendees, and the festival was conferring awards second to none in prestige, nurturing talent with radio residencies at a retreat on the outskirts of Chicago and short doc competitions, while generating a vital online archive of twenty-first century creative radio storytelling. Third Coast meetings aired deep problems in the industry, fostered new projects, and oddly coincided with national shocks—the first conference took place right after 9/11, another meeting just days after the Trump election. It also featured speakers who bridged radio and podcasting, such as Jad Abumrad of *Radiolab*, Ira Glass of *This American Life*, Joe Richman of *Radio Diaries*, legends like Jay Allison and The Kitchen Sisters, as well as many figures discussed in the previous chapters of this book. On the same day as the "Revolution" panel, another speaker was Alix Spiegel, a *This American Life* producer who had met erstwhile *Radiolab* producer Lulu Miller at a previous Third Coast. That year the pair launched *Invisibilia*. Phoebe Judge's *Criminal*, the gold standard for a quality crime podcast, won the Best Documentary award that night for "695BGK," which told the story of how a police error in inputting a license plate led to the shooting of an unarmed Black man. The keynote was *Mystery Show* host Starlee Kine, fresh off an appearance on Conan O'Brien's TV show and the very embodiment of a flash-in-the-pan podcast star, bringing the house down with a talk in which the boisterous crowd gave her pitches for her show.

Third Coast's turn in the 2010s from a family reunion into a recruiting scene, with exhibit tables from Audible, the *New York Times*, Descript, and others, was a signal of the arrival of a giddy proleptic imaginary in the podcast space. The personnel of other conferences studied Third Coast. In 2015 I met a founder of Podcast Movement at a Third Coast event; in

2019 I met a founder of Sound Education there. Attending in 2017, *New Yorker* writer Sarah Larson opined that, thanks to podcasting, what had once been a low-key liberal media get-together had really turned into the "Wild West," language I discussed at the outset of this book as redolent of podcasting's jejune era of next-big-thingness.[6] In the 2010s there would be many other venues for podcast fiction people to meet. But Seattle's alt-culture-oriented Podcon would not become a nexus for indie creators until 2017, and neither Podcast Movement nor the London Podcast Festival would have a fiction track until 2019. So Heppermann and Johnson's proposal arrived at a time and place at which it was not yet clear if podcasting could do for fiction what it did for true crime, taking it out of the black box and onto the red carpet. Most attendees were journalists by profession; historically, radio dramatists typically trained as directors, composers, writers, or poets. Even if they wanted to try fiction, there were few models. While more and more producers were spending time on stages by then, with tours of *The Moth*, *Snap Judgment*, and *Radiolab*, many in the field still did not even know what to call audio drama. At this event I heard terms like "podcast fiction," "scripted series," and "non-non-fiction."

It is striking that three of the pieces that defined what counted as "non-non-fiction" around 2015—"Tape Delay," "Movies," and "Every Heart"—use an identical pretext, the encounter of two strangers in New York, as a canvas for techniques that access interior impulses presented to us as only expressible in audio media, the tacit criterion that made them "audio fiction" rather than stagey "radio drama" or sci-fi-redolent "audio drama." In each case the nominal story is not in the spotlight. The pieces are metasonic narratives, in that they reflexively reinscribe across their events and emotional moments the very digital audio tools that make possible their production, to the extent that, although these plays were not stealth marketing for high-end DAWs, they could have been. The "idea" of the work is in the editing or, more precisely, in the conjoined notions that layered, interiorized vocal consciousness, whether jilted, manic, or reflective, is expressible through the tracks of a DAW session and that audio editing is at its heart a form of experimental literary writing. Even the protagonists in these stories can be understood as doing a kind of sound editing (psychic or material) across the arcs of these narratives. This makes these pieces irreducibly sonic. Transcribe them to text and they flatten to nothing; listen to them without attending to their use of stereo and their structure bleeds away. It is not clear you can really experience them without high-quality headphones. Audio editing is both theme and practice in these works, so listeners can be forgiven for forgetting the characters, settings, and actions. In this moment, to be "revolutionary," audio fiction had to crystalize as a method of experience rather than as a genre of story or set of thematic stances.

Vividly displaying these plays and their common approaches, "The Revolution Will Not Be Televised" impressed attendees with the idea that a brand-new form of narrative audio was out there, waiting to be found. Whether the work that was being described was truly new and whether it would yield works of durable value are complicated topics I explore in some detail below, but for the moment I want to emphasize that coming years would see this exact appeal repeated more or less constantly in a wide variety of settings, as rewound and repeated as the narrative threads in "Tape Delay," "Movies," and "Every Heart." While the Third Coast session stands out as particularly formalistic in its approach—most other subsequent direct efforts at "inventing" a new form of audio fiction storytelling for the podcast era would focus on writing, fan communities, characters, and plots—from the mid-2010s down to the present, the making of fictional podcasts would be consistently portrayed as emergent and fresh, about to break through into the center of popular culture, waiting for its "*Serial* moment."[7] The heyday of podcast audio fiction seemed just around the corner, year in and year out, a situation that at once shows just how powerful the rhetoric of next-big-thingness was at the time but also, and more fundamentally, provided the genre with its most basic poetic precondition.

The Aesthetics of Amnesia

This chapter explores the world of fiction-based dramatic podcasting in the late 2010s, a period that saw compelling productions at all levels, from independent creators releasing whole seasons of pieces cut from whole cloth to full-cast adaptation of epics with famous actors and impressive budgets. In another way, the story of this period is strikingly familiar. In chronicling some of the works of these years, I point to how the rise of audio drama in the 2010s was just one of the many "rebirths" of this kind of storytelling in its century of history, arguing that soundwork has long been discursively characterized in a paradoxical manner—as a medium thoroughly of the past but also as a medium without a past, one whose previous iterations were so lacking in merit as to be best forgotten entirely, a medium that is always just about to be "really" born—by those who feel they finally have the tools to do so. In the 2010s, many creators yearned for an audio drama that sounded "modern," perhaps unaware that prominent dramatists had articulated a virtually identical thought in the decades before them: the *Columbia Workshop* writers in the 1930s; Richard Durham's *Destination Freedom* in the 1940s; the writers of the British golden age in the 1950s; the artists of the Neues Hörspiel movement in Germany in the 1960s; *The National Radio Theater of Chicago* in the 1970s; and a host of until recently unsung radio creators across the Global South from Severo Sarduy of Cuba to Alexius Buthelezi of South Africa.[8] Audio drama has been made "new" many times; indeed, its hallmark has been its perpetual availability to be rescued from its own alleged backwardness.

As in previous chapters, here I approach new audio fiction broadly and closely, looking at a few key trends across programs, and analyze specific shows whose styles are characteristic of the period, such as *Serendipity* and *Limetown*, while offering detailed analysis of standout shows dealing with memory at the aesthetic level that I think likely to be permanent additions to audio drama history. I will show that audio fiction was not "revolutionized" in the period—there is as strong a case to be made for incremental progress as there is for sudden change—but that, instead, dramatists for various reasons successfully produced a discourse about its imminent reinvention, one that took place in meetings and commentaries, in reviews and edits, and that we can also hear in the work itself. Like many predecessors in the history of narrative audio, podcasters felt that a moment of rich ambiguity had arrived when what would count as audio drama could be redefined. This change justified artistic experiments that seemed fresh (whether or not they truly were), and their investments stood as yet more evidence that a moment of medium ambiguity had arrived in which a transformation was just about to happen.

It takes little effort to show audio drama as a field preoccupied in this period with what it meant to forget and to remember. Listen to podcasts of the late 2010s for any duration and you will find many characters and plotlines from *Limetown* to *The Far Meridian* and *Unwell* marked by memory impairment, mismemory, or poisoned memory in a way that structures the story and how it is narrated. The prevalence of psychiatric sessions aimed at memory retrieval in serials from *The Bright Sessions* and *Homecoming* to *The Ghost Tape* and *Batman: Unburied* is another sign of this fascination, as is the prevalence of many female hero reporters—Lia Haddock in *Limetown*, Nicki Tomalin in *The Message*—who turn out to be ciphers, even to themselves. Audio drama is not unusual in using amnesia stories or memory-recovery frames. As David Bordwell has pointed out, "magical forgetting" has long been reliable as a device to open possibilities for a whole set of desirable narrative and emotional effects, including comedy and suspense, recognition, sympathy, and flashback.[9] But audio dramas of this period also took further steps. Many speculative fiction serials take as their backdrop the erasure or removal of information technology—*Bellweather*, *Blackout*, *Case 63*—usually in the form of speculating on some future destruction of the internet, while others concretize concepts of fragmented memory by thematically exploring recording itself. Tape, the one thing digital audio dramas did not need, is everywhere in audio drama stories of this period, from true independent shows like *Mabel* to network scripted series like *Within the Wires* and celebrity-driven pieces like *Blackout*. In *Archive 81* we hear podcaster Mark Sollinger listening to tapes of recently disappeared archivist Dan Powell, who himself was listening to tapes of mysterious urban researcher Melody Pendras. The long-running UK serial *The Magnus Archives* starts with a similar premise—an archivist following another archivist. *Video Palace* told the story of Mark Cambria, a collector of obscure video tapes whose mind is infected by a near-mythical white VHS videotape, which makes him speak in tongues as he sleeps and takes him on a journey into tape collecting subcultures. In *Darkest Night*, two researchers working for a dystopian corporation dissect the eyeballs of the recently deceased to access their stored memories, as if these body parts were themselves video tapes, a squelchy anthology recorded in 3D audio so that each time an eyeball is accessed it feels like it's our own. It is remarkable that these dark fantasies and horror shows share the very same fascination with memory and the same poetic ruminations over material recording—including the actual use of analogue recording device sounds—as prestige shows like "Tape Delay," "Movies," and "Every Heart," suggesting some common bond. Whether it was through characters, backstory elements, props, or sonic textures, many audio fiction podcast creators were clearly trying to work out something about forgetting, recording, and remembering.

This thematic fascination also begs the question: did podcast fiction itself, as an art form, bear memory of its ancestry in radio fiction, the kind of thing we tend to mark when we catch a habit of homage or citation? Andrew Bottomley has made the case that *Night Vale* stands as a "remediation" of radio drama, and there is certainly much truth to this.[10] There are also a few shows that are more direct in their references, as in an allusion to Orson Welles's 1938 *War of the Worlds* at the outset of *We're Alive* and the use of the faux-noir dramas that animate *What's the Frequency?* Yet this period shows little to no work that pays explicit homage to the audio dramas of most of the writers associated with the medium over time: Carl Sandburg, Edna St. Vincent Millay, Louis MacNeice, D. G. Bridson, Caryl Churchill, Arthur Kopit, Sylvia Plath, Samuel Beckett, Norman Corwin, Langston Hughes, Dorothy L. Sayers, Douglas Adams, Erik Bauersfeld, Bertolt Brecht, Tom Stoppard, Arch Oboler, Harold Pinter, and others. This fascination with memory paired with skepticism toward homage gives many of the great pieces of this period their unique feel. *Homecoming* and *The Shadows*, the two pieces that this chapter eventually arrives at, are each in their own way exemplary of a way of "doing" a memory play without necessarily having an explicit memory of their generic roots. What's more, in these and many other cases, historical precedents lurk behind these works anyway, like involuntary recollections. In the attempt to move away from old radio drama, evading it through a theater of forgetfulness, fiction podcasting could even redeem the very predecessor it tried to surpass. To borrow a term from philosopher Jacques Derrida, this was a period in which audio drama found itself "*en mal d'archive*" in a triple sense: it told stories of memory sickness, it was emphatically sick of its own archive, and yet at the same time it felt an underlying itch for what the archive offered.[11] Radio drama was audio drama's shadow, an adumbration that shaped objects by sharpening their outline, vanishing almost behind them, and sometimes reappearing.

To identify the way the medium dealt with forgetting and remembering, I would like to employ as a descriptor the *aesthetics of amnesia*, a name for the shifting issue of recall in the genre and its works in the period. The aesthetics of amnesia is intended as a framework for conceptualizing podcast-era audio drama that seems bound up in questions of recollection in a creative field whose leaders often showed disdain for, or pointed naiveté toward, the radiophonic roots of the medium, acting like audio drama is completely new, while at the same time using ideas and techniques that functionally resurrect the history of the medium.

Even this amnesia *itself* is not new. Amnesiac aesthetics tend to appear at moments in audio drama history when technological change is perceived to be frictive, generative, lucrative, or all three. Uniquely among arts, radio drama has in its long history developed no habit of homage, what Michele Hilmes calls "a sense of expressive continuity" that links one work to the next, and so creators tend not to obsessively explore antecedents with the vigor that filmmakers study films, novelists novels, or poets poems, rarely producing works of radio *about* other works of radio.[12] Indeed, today's podcasts are striking for the *lack* of perceivable "anxiety of influence," as critic Harold Bloom once described the literary disposition among canonical Western writers of poetry. Bloom famously characterizes the author as a passionate figure who falls in love with their preceding textual influences but also suffers from the presence of those influences, undertaking a strong but ultimately partial reading of works of the past—he calls it "poetic misprision"—whose consequence is the work the artist finally produces. There is a certain appeal to the melodrama of this process, though it is also easy to see it as highly conservative and exclusionary of outsider perspectives. At any rate, the model cannot be extended to radio drama as a field, in part because until very recently it was difficult to obtain historical pieces, so the initial process of "strong reading" does not take place. There is no *Norton Anthology* of radio drama, disenabling the route by which poetic misprision is supposed to happen. In its place, a culture has arisen around the form that tends to see neither requirement nor reward for emphatic contextual placement in radiophonic tradition, an aesthetics of amnesia.

This is not to say that artists had no context for fiction. Many podcasters took inspiration from genre film and television programs, as well as work created for the theater, fields in which many did (or aspired to) work. This was particularly true for professional journalists dabbling in fiction. At a Third Coast panel in 2019, for example, I attended a talk by producers Arwen Nicks of KPCC's *The Big One*, Sam Greenspan of *Bellweather*, and Zoha Zokaei of *The Price of Secrecy*, who as a group were discussing their inventive and compelling uses of fictional elements in high-profile fact-based stories. When conversation turned to citing their influences, the panelists had a lengthy discussion about long-form reporting and canonical nonfiction, New Journalism classics and science fiction novels, prestige TV and reality shows, Iranian social realist films, and neuroscience texts, but not a single radio drama. And this despite the fact that the blending of fiction and nonfiction has produced the most memorable and best-preserved broadcasts of all time, from Archibald MacLeish's *The Fall of the City* and Orson Welles's *The War of the Worlds* in the 1930s to the radio features of Piers Plowright in the 1980s and 1990s. Moreover, whereas previous generations might have been unable to access more obscure titles, the advent of the MP3 meant that organizations from OTRCAT to Archive.org hosted hundreds of thousands of examples of classic radio for dramatists to study, were they interested in nesting new work using the kind of contextualization rewarded in other arts. The history of audio drama is more researchable now than at any time in history but remains just as ignored as ever.

Is this disinterest in self-contextualization a problem? The reflexive feeling that it is wrong to reinvent without remembering is itself a construction, coming from a tendency to think of memory as a straightforward circuitry in which any failed recollection is coded as a repression or pathology. Media theory itself has long been entangled with this idea. Indeed, when theorists and historians have approached the concept of amnesia, it has often been to show how narratives expose a particularly modern model of consciousness, memory, and recollection, a tendency that resonates across several otherwise divergent areas in media studies. Thomas Elsaesser, for example, notes that amnesia in narrative is linked to the replacement of "remembering" with "programming" in contemporary films, while Gilles Deleuze associates the cinematic history of amnesia more closely with "dream" than with "recollection."[13] In both cases, theory of amnesia begins as a story about how trauma produces divergence from a preexisting circuit around which the relations between perception and memory are normatively regulated (this is rooted in Freud's model of consciousness in *Beyond the Pleasure Principle*), something that cinema narrative has a penchant for exploring and that critics have a similar yen to map.[14] It is tempting to go down this road all the way and to see the behavior of modern audio drama as a straightforward instance of Freud's idea that the history we repress becomes a behavior we repeat.

But the route of pathology offers little benefit to this project, nor does an effort to create and defend a canon to which creators must defer. After all, when podcasters fail to cite their ancestors in radio drama, is that really an infelicitous situation? Bloom writes of how one result of the anxiety of influence is that the voices of the forerunners come back through the misprision of successors, and it would be a strange argument to insist that audio dramatists must recapitulate this ghoulish patriarchal ritual just to be taken seriously. The chain of poetic misprision has not always yielded in literature the diverse population of creators that audio drama has in recent years been able to foster, including members of communities of color and gender nonconforming individuals. If the aesthetics of amnesia is the opposite of poetic misprision, which would account for the odd "perpetual present" of the form, then perhaps this is a strength rather than a weakness. Forgetfulness opens space even as it forecloses it, and many narratives feature a kind of amnesia that is just as generative, restorative, and creative as it is problematic. If critic Andreas Huyssen is right, that modern societies are today undergoing a panic over amnesia that produces a "memory boom" in search of cultural works that offer anamnestic rituals of remembering, audio drama is an especially valuable form for both sides of the process, as it embraces both forgetting and remembering at once.[15] On one level, then, a focus on the aesthetics of amnesia asks how audio dramas come to remember and how they come to forget, while on another level it asks what becomes possible when doing so. At its best, the aesthetics of amnesia could be a way of coming to grips with audio drama's knotted relationship with memory culture, with many of the most thoughtful creators using the medium to investigate the nature of nostalgic affect.

With all this in mind, my focus here is on how the aesthetics of amnesia were dramatized and discursively produced and understood. Audio dramas that showcase self-divestment from memory are *situated* that way by contextual elements, and that situatedness also permits (even solicits) contemplation about how they could be framed otherwise, and one task of the critic is to speculate on that, asking what history a particular piece might be attempting to subsume. Criticism that comes from a place of archival memory is an important theme of this chapter, and one thing I explore is how amnesiac features of these stories often point toward, and are vivified by, performances of their own removal within a piece. I therefore intend the aesthetics of amnesia to refer both to the formal *state* of not remembering (like the archetypal amnesia sufferer who has suffered a blow to the head) and to the *process* of coming-to-remember by which that initial state is illuminated (the archetypal second blow by which memory is restored).[16] It is not merely that "audio drama" replaced "radio drama" but that it did so in a kind of vicious loop in which an unawareness of the past of the form was replaced by awareness and then forgetting again. As a result, this chapter captures a double movement in the mid-2010s, showing how the medium's process of "revolution" both prompted and sublimated a parallel but equally powerful process of recall. Podcasts produced in the 2014–20 period relate to the ones that came before, even if they do not overtly recognize it, just as those made in the past likewise owe their own unacknowledged predecessors, all the way back.

A Naming Crisis

The term "audio drama" predates the term "podcast." Radio scholar and historian Tim Crook used it extensively in a key 1999 book that frequently adopts a "radio and audio drama" phrasing and associates "audio drama" both with theater prior to the age of radio and with more recent online work. For Crook there is no clear "radio drama" *then* "audio drama" sequence of events, nor is there a perfect distinction between the two, though he did presciently see creators entering the "sixth" age of audio drama today, one focused on internet-based production rather than terrestrial radio.[17] The first usage of "audio drama" in the current sense in the popular press also came in the previous century, in a 1998 *New York Times* article focused on webcasts, live performance, and books on cassette and CD; the term only later evolved to indicate direct-to-digital works that are materially distinct from "born-radio" dramas like *L.A. Theatre Works* or dramas from public broadcasters and amateurs.[18] Other terms from texts and articles of that time include "scripted series" and "full-cast audiobook," terms used in public radio and audiobook communities respectively. When independent creators in the mid-2010s began to coalesce around the term "audio drama" to make their work searchable to fans and peers, as critic Wil Williams has explained, it was partly as a nod to "radio drama" as its roots.[19]

It was against this backdrop that another trend began for "audio fiction," a preferred term at the Sarah Awards mentioned above. The jockeying among these various terms briefly came to a stop in the summer of 2019, when "fiction podcast" became a term adopted by Apple, whose IOS app created a category for this type of work for the first time, a development that was seen as both an affirmative recognition for a medium in which many had toiled for years and a corporate appropriation. Now Apple's monopoly on terminology seems to have waned, and it is not clear that terminology has come to any state of rest. Spotify, today's busiest podcast interface, classifies most audio fiction under the global category "stories," which also includes nonfiction narrative podcasts of various kinds, subcategorizing fiction by genre rather than mode.

The lack of consensus during these years around terminology suggests several things. For one, practitioners clearly had an elastic relationship with other media and with the rootedness of their practice in radio. For another, the onslaught of new terminology and concomitant inability to name the practice definitively had a relationship with productivity in the field. As D. N. Rodowick has observed in the case of cinema, a naming crisis can be as generative as it is nettlesome, and it occurs when a medium finds itself in a state of undecidability about both its relation to its genealogy and its prospective future.[20] Indeed, it was because space was opened by a naming crisis, and not in spite of it, that an aesthetic of amnesia could develop. For the purposes of this chapter, I want to honor the importance of the naming crisis by keeping these terms a little fluid and to emphasize lines of convergence, but it is helpful to distinguish between four communities or types of audio production in order to map out groups of productions that I heard in the period. The first is the "audio fiction" tradition represented by *Serendipity*, *The Truth*, and other creative works cited in the last section, the sort that generally had experimental sound design, appeared in one-offs and anthology shows rather than serials, and tended to appeal to Third Coast–style audiences. In addition to that, there were at least three other types of audio drama-making operating in the late 2010s: independent productions (usually "audio drama"), works emerging from the audiobook world (usually "full cast audiobooks"), and works emerging from the slate of new podcast networks (variously described as "audio drama," "podcast fiction," or "scripted series").

I dealt with the first category in the opening of the chapter, so what about the second? In 2015, under the ears of most podcasters at events like Third Coast, independent work in audio drama was already a fully formed online subculture. That year a search for "audio drama" on Google brought back enthusiast listservs and directories of programs in MP3 format that ran into the hundreds, then thousands, available on iTunes accounts, Soundcloud pages, and YouTube channels. Many were made by amateurs and small production companies inspired by long-running self-distributed serials of postapocalyptic survival such as Christof Laputka's *The Leviathan Chronicles* (2008–present), Kc Wayland's *We're Alive* (2009–16), and Fred Greenhalgh's *The Cleansed* (2013–19) that had cult fan bases (according to their publicity, *We're Alive* has been downloaded fifty million times). All were serials, but rather than giving us a drama of sequential episodes in a regular time slot, these programs put their episodic open-ended structure atop an on-demand media practice. In 2015 newer plays included a wide variety of scales of production: limited works, such as Rick Coste's *The Behemoth*, about a mysterious giant walking across Massachusetts with a young girl; homages to noir and folk horror, like James Urbaniak's Hollywood mystery *A Night Called Tomorrow* and John Ballentine's anthology *Campfire Radio Theater*; and sprawling narratives that take place in speculative future dystopias, alternative presents, and distant futures, often after colossal disasters, like *Our Fair City*, *Transmissions from Colony One*, *Wolf 359*, *Tin Can*, *The Bridge*, and *The Orphans*.

It was an incredibly productive, and largely unremunerated, community. That fall, while experimental shows were drawing small festival crowds at Third Coast, a serial called *The Black Tapes* by Terry Miles and Paul Bae was drawing up to two hundred thousand listeners a month. The series featured radio reporter Alex Reagan and skeptical scientist Richard Strand on the trail of geometry cults whose demonic doings were hinted at in the titular videotapes. With stories of possessions, upside-down faces, and a sinister order of Russian monks, *The Black Tapes* spawned two companion pieces—*Rabbits*, about a centuries-old live-action role-playing game with deadly consequences, and *TANIS*, about a public radio host in search of an ancient legend—as well as a host of imitators. Bae would go on to produce *The Big Loop*, an anthology program with "stories of finite beings in an infinite universe," ranging from a monologue of supernatural hauntings in a dance studio to the stories of professional grief surrogates.

The previous summer *The Message* had also been released, a narrative about podcaster Nicky Tomalin working with a team of codebreakers to decipher an alien message, as a deadly pulmonary virus spread around the world, seemingly borne by the sound itself. The program showed how scaled narrative could be narrowed to the pinhole of a microphone, modeling low-budget high-concept work. Focusing on a reporter with a secret and a series of archival tapes, *The Message* set many of the conventions that would later become cliché. Its contributors would also go on to make work that appealed to a wide variety of backers. Radio veteran John Dryden, who produced the piece, would soon be making the *Game of Thrones*–style serial *Tumanbay* for BBC, which was redistributed in podcast form in the United States by Panoply. *The Message* sound designer Brendan Baker would cowrite and direct long-form *Wolverine* stories for Marvel. Playwright Mac Rogers would write a sequel to *The Message* called *Lif-e.af/ter*, a Gibsonian piece about a lowly FBI flack obsessed with speaking to his dead wife, her voice made immortal in an audio social media ecosystem that seems to have its own agenda and awareness.[21]

The new serials also expanded personnel, and with production scope came larger narrative worlds. *The Message* listed a cast and crew of about twenty people, a number that was doubled to create *Lif-e.af/ter*. Rogers would also write the memorable serial *Steal the Stars*, a brilliantly directed sci-fi thriller about two elite soldiers who fall in love while working at a secret facility with an alien in a coma. Taken as a whole, the indie genre had a fairly clear set of influences: H. P. Lovecraft and Philip K. Dick; *Twin Peaks*, *Black Mirror*, and *Lost*; *The X-Files* and *Star Trek*; and *The Blair Witch Project*, *Paranormal Activity*, and other found footage films. Martin Spinelli and Lance Dann have observed that a show such as *The Black Tapes* "taps into the audience's collective memory with scenes from the canon of the last twenty-five years of screen horror."[22]

Perhaps the best-known drama in the indie category was then, and still is, *Welcome to Night Vale*. Joseph Fink's long-running series was already in its third year by 2015, featuring a droll community broadcaster named Cecil blithely giving the local news in a small desert town subject to a glow cloud and any number of dark government conspiracies and paranormal forces, while mooning over his love Carlos. The show manages extraordinary world building, using few tools, along with a strong sense of farce. *Night Vale* was clearly a disruptor when it came to attracting podcast listeners to fiction and, as Ella Watts has shown, was justly celebrated by its fans for the remarkable diversity of its characters along lines of gender, sexuality, race, and ability.[23] *Night Vale* also has an interesting minimalism. Its production values pale next to those of contemporaneous audio fiction, from *The Truth*'s postproduction-forward philosophical fables to the gunslinger tales of Random House's *Louis L'Amour* full cast books, some of which took years to produce. But *Night Vale* was surely successful at leveraging its popularity. In 2015 it went from a podcast and touring show to a veritable brand, announcing plans for a new series entitled *Alice Isn't Dead*, about a trucker named Keisha searching the country for her long-lost wife, all the while pursued by a hideous "Thistle Man" terrorizing her on the highway, a series that consists of monologues punctuated by the sound of a CB radio's switch. *Alice* would be followed by *Within the Wires*, set in an alternate history that takes place in the margins of a set of relaxation tapes bookended with sounds. In these cases, as Watts points out, shows put characters who identify as queer, disabled, or marginalized at the center of indie programs from *Night Vale* to *The Magnus Archives*, which resulted in much artistic growth in the 2010s.

If *Night Vale* was the mainstay in 2015, the newcomer was *Limetown*, an independent serial from Two-Up productions about a mysterious mass disappearance in a Tennessee research town and the reporter investigating it years later, a fiction podcast that perhaps better than any other demonstrates Leslie McMurtry's argument that audio dramas took on "centripetal intimacy" along with seriality as a hallmark of post-*Serial* podcast aesthetics.[24] With twenty-seven actors and a sound design that drastically excelled the minimalist approach, *Limetown* made it to the top of iTunes within three weeks of launch and was one of several podcasts to blur the line between independent and mainstream. Either way, a robust online community formed around independent pieces, often coming together in how-to podcasts like *Audio Drama Production Podcast* (begun in 2013) and *Radio Drama Revival* (begun in 2007), as well as review websites like *The Timbre*,

Hot Pod, and eventually Bello Collective, where creators were interviewed and new projects lauded. In these venues you could catch a dedicated, scrappy community focused on how to attract sponsors, coach actors, set up mics, and build a page for contributions on Patreon or other sites, all while hearing creators geek out about one another's programs about zombies and spies, robots and spirits, horror and porn, conspiracy and fantasy. Creators from the independent world sometimes crossed over into larger better-financed projects while often working hard to keep the values of the community alive. Fred Greenhalgh of *The Cleansed* later worked for high-profile Audible *X-Files* projects and often offered free online classes and tutorials that inspired many emerging creators; Kc Wayland of *We're Alive* made the Lawrence Fishburne vehicle *Bronzeville* and self-published a definitive and understudied text about how to make modern audio fiction.[25] Today, many of the people who work as advocates of (and watchdogs for) the genre came from this independent world. Creators Wil Williams of *Valence* and Tal Minear of *Someone Dies in this Elevator*, in addition to working on their own pieces, recently held audio fiction company Rusty Quill to account by supporting journalism critical of that company's practices and distributed newsletters highlighting the work of trans and non-binary creators.[26]

In parallel with the prestige audio fiction track and the broad lane of independents, a third stream of creative audio work ran as well, joining the other two from an older "full-cast audiobook" genre that had been developed in its modern form by Yuri Rasovsky, director of *The National Radio Theater of Chicago* and later of *Hollywood Theater of the Ear* and a prolific audiobook creator for Hachette Audio, with works like his adaptations of *Sweeney Todd* (2007), *The Maltese Falcon* (2008), and *The Cabinet of Dr. Caligari* (1998) that bridged the age of the CD and the age of online streaming. Rasovsky passed away in 2012, but his influence could be felt by the time Amazon subsidiary Audible began some of its first experiments in the genre two or three years later, such as Jeffrey Deaver's *The Starling Project*, Sebastian Fitzek's *The Child*, Joe Hill's *Locke & Key*, and Dirk Maggs and Tim Lebbon's *Alien: Out of the Shadows*, which was followed up by two more Alien franchise pieces. Using these vehicles, Audible earned press in the *New York Times* and dominated the "full cast audiobook" genre, which would soon become a whole business model.[27] Audible's platform supported independent successes in the field such as Bafflegab's radio adaptation of the folk horror *Blood on Satan's Claw*, which won a New York Festivals award in 2018, as well as a number of odd experiments that are not well remembered enough, such as the full cast reading of George Saunders's 2017 experimental novel *Lincoln in the Bardo* with some fourteen celebrities, including David Sedaris, Don Cheadle, Julianne Moore, Nick Offerman, Keegan-Michael Key, and Susan Sarandon.

This was also the period of several major BBC adaptations of Neil Gaiman's books—*Neverwhere* (2013), *Good Omens* (2015), and *Anansi Boys* (2017)—that starred the likes of James McEvoy, Benedict Cumberbatch, and Lenny Henry and would eventually find their way into the Audible universe. All of them were adapted by Dirk Maggs, sometimes with the direction of radio drama veteran Heather Larmour, and reflected Maggs's effortlessly direct style: clean vocals, long-form stories, bright and clear sound effects, fast action, and lush but articulate and eminently legible broadband backgrounds. The musical Maggs style, which you can hear in his *Hitchhiker's Guide to the Galaxy* work and many comic book adaptations series before coming to the podcast world, remains a gold standard in the field.[28] The relatively narrow genre scope of "full cast audio drama" works—fantasy and horror, nostalgia genre fiction, bought-at-the-airport thrillers—would change in coming years. In 2017 Audible began an Audible Theater division under producer Kate Navin with a $5 million fund that sought new scripts to be juried by the likes of David Henry Hwang, Tom Stoppard, and Annette Bening. Soon it would be producing stage plays, like Chris Kelly's "Girls and Boys" starring Carey Mulligan, and releasing them as audiobooks. By 2019, with new shows on stages off-Broadway and inside recording studios, the *Washington Post* would be describing Audible as a "digital repertory company" with platforms "on air and on legs."[29]

Audible is an idiosyncratic brand in the new audio industry, neither a true legacy house like NPR and the BBC nor a brazen upstart company native to the podcasting space. The latter category—brands such as Panoply, Howl, Wondery, Gimlet, and Radiotopia—was the source for perhaps the most high-profile audio fiction and scripted series to emerge in the late 2010s in terms of publicity, and this constitutes the last of the four types of audio production: audio dramas developed in the context of a larger slate of nonfiction offerings by born-podcast companies. That trend began with stage-based audio plays in the comedy genre that typically included improvised characters and scenes very early on in the podcast era, including *The Thrilling Adventure Hour* and *Comedy Bang! Bang!* It also emerged from the use of audio drama in vignettes in true crime programs—*Crimetown*, *Stranglers*, *Unsolved Murders*, *Hollywood and Crime*, *Inside Psycho*—that employ dramatic re-creations to an extent you don't hear in more professional investigation-based programs like *Serial* and *In the Dark*. Vignette drama could take a number of forms, ranging from a small illustration in a scene to a playlet. For example, a show like *Stranglers* had typewriter sounds subtly illustrating the work of a reporter whose pursuits were being narrated, while *Crimetown* might use footsteps, car sounds, and doors in the background to illustrate a particularly harrowing tale of a night in the life of the Providence mafia. *Hollywood and Crime* used re-creations of scenes from the infamous 1947 Black Dahlia case that can take up the foreground entirely in a way that undermines their seriousness, including scenes of reporter scrums down at headquarters, with on-the-nose noir patter.

Like the "full cast audiobook" category, born-podcast company "scripted series" pieces for new networks often featured celebrities: Jenna Elfman in *Secrets, Crimes & Audiotape* beginning in 2016 with Wondery; David Schwimmer, Oscar Isaac, and Catherine Keener in the 2016 thriller *Homecoming* with Gimlet; and Jemaine Clement in the farcical Monty Pythonesque adventure *The Mysterious Secrets of Uncle Bertie's Botanarium* for Howl in 2017. Soon you could hear Kristen Wiig and Ethan Hawke in dramas on some of those same networks, and by 2019 all the prominent content houses trotted out talent regularly. Rami Malek's series *Blackout* for Endeavor Audio and QCode came out just after he won an Academy Award, and the company followed up with *Hunted*, a US marshals thriller produced by TV legend Dick Wolf starring indie film mainstay Parker Posey. In 2019, rapper Method Man appeared on *Marvels*, an expansive project for Stitcher set in the Marvel Universe, while musical theater actor John Cameron Mitchell began work on an experimental podcast musical, *Anthem*. Radiotopia, arguably the leading network in terms of artistry in all of podcasting and long the home of *The Truth*, began its first serial fiction that year, too, *The Passenger List*, with Kelly Marie Tran and Broadway legend Patti LuPone. These programs used more resources than independents or full cast audiobooks, and even those that were the most independent had a Hollywood production style. Lawrence Fishburne's historical drama *Bronzeville*, which focuses on the lives and loves of African Americans in Chicago's famed nightclub district in the 1940s, lists more than fifty members of its cast, crew, and staff in the first episode. They even had a caterer.

Podcast network-style pieces attracted the most press, if not always the most listeners. They had a capacity to react to trends, build communities, and push boundaries. Just three months after the program *Stranger Things* was a TV hit for Netflix in 2016, Panoply had a scripted series for children that had similar themes and terrific acting, *The Unexplainable Disappearance of Mars Patel*, about a group of kids on the cusp of adolescence who fall upon a kidnapping conspiracy at their school and are soon on an interplanetary adventure. *Mars Patel* is distinguished by the use of an internal character narrator named Oliver Pruitt, who seems to be the principal villain of the series and also presents himself as its sponsor. In 2017 *Mars Patel* was the first born-podcast audio drama to win a Peabody Award, in fact the first audio drama of any kind to win since 2009. Podcasts also branched out into other genres and publics. The British sitcom *Wooden Overcoats*, which has attracted a transatlantic following, is the comic story of a mad rivalry between two funeral directors on a small English island, punctuated by perceived slights, unrequited love, betrayal, and elaborate plots, often told from the perspective of a pet mouse. In 2016 an exciting piece was the thriller *Fruit* by indie TV wunderkind Issa Rae for Midroll media. The suspense series adopts first-person retrospective narration, one of the forms that radio itself helped to popularize in American fiction generations ago but which has since been difficult to reintroduce among dramatists who prefer the present tense. Taking the form of a monologue that segues into extensive scenes from the recent past, *Fruit* is narrated by X, a young pro football player, to a hidden narratee at some point in the future. The plot begins when X locks eyes with another man across a crowded club, and we know that his sexual orientation is awakening. The series then follows X's growing awareness of his own desires in a homophobic professional context. *Fruit* is a blend of corporate thriller, sports story, and erotica that pushes all three genres, including surprisingly explicit depictions of sex. At the time of writing, the season was moved behind a paywall.

A Medium without a Memory

With this sketch of the landscape of "non-non-narrative podcast fiction," "audio drama," "full cast audiobooks," and "scripted series" in mind, I would like to turn to how these pieces were talked about by critics and practitioners in the late 2010s. Although these works had a hard time naming themselves as a group, one thing they shared was a predilection for self-defining against "radio drama," a term that itself has long been contested and still is today, as Leslie McMurtry has recently pointed out.[30] Two decades ago, when I began researching this form, I opted to follow the minimalist idea that radio drama should be on-air and ought to feature a "scene" in which a "character" performs an "action."[31] This was my way of moving away from the tendency for audio dramatists to speak of their craft somewhat simplistically as an amalgam of music, words, sound effects, and silence, which struck me as true but not helpful—a little like a writer saying that the way to write is to use adjectives, nouns, verbs, adverbs, and punctuation.[32] This definition was also a way of distinguishing audio drama from reading aloud, which shares many of its features with audio drama, such as speaking "in character" and frequent use of musical score or even sound effects, but tends to lack depicted action and scenography as understood from an audioposition, which is why many people intuitively consider audiobooks to be a different narrative audio mode.[33]

But setting out even minimalist parameters like those does nothing to restrict real usage of the term. Follow the natural uses of "radio play" over the last nine decades and you will find that it is neither always "radio" (it also refers perfectly regularly to audio on vinyl and tapes of radio exchanged among collectors and stored in formal archives; "liveness" in production and reception is an exception, not the rule) nor always "drama" or a "play" (it can lack story, action, characters, or settings). Even within the context of the podcasting boom today, a lot of exciting material associated with the radio play genre is unavailable on air or online, appearing instead as a CD (*Dark Adventure Radio Theatre* is an example) or only onstage (Chicago's Wildclaw Theatre for more than a decade held worldwide competitions for onstage "radio" horror plays complete with live effects by incomparable foley artist Ele Matelan). At the same time, the relation of audio to the sort of drama associated with theater is unclear. As literary critic Milton Kaplan once observed, many radio plays did not work on stage because they tend to have very little conflict, so while theater pieces are sometimes remade for radio, few radio pieces are remade for stage.[34] Often in audio the tie to theater is abandoned in favor of the metaphor of cinema, something that speaks to varied audiences—on the Radiotopia network, *The Truth* advertises its complex, artful eighteen-minute experimental vignettes as "movies for your ears," while in the CD section of truck stops Graphic Audio calls its full cast epic adaptation series (*A Court of Thorns and Roses* by Sarah J. Maas, 46 hours; *The First Mountain Man* by William W. Johnstone, 149 hours; *The Stormlight Archive* by Brandon Sanderson, 164 hours) "movies for your mind."

As these last examples emphasize, setting out a sense of scale and theme for the genre is challenging. It is hard to imagine a category that contains both the workhorse dramatic BBC soap serial *The Archers*, a series about "everyday country folk" that has been running for some seventy years and is nineteen thousand episodes long, and Antonin Artaud's 1947 scatological indictment of modernity, *To Have Done with the Judgment of God*, a good portion of which is terrifying tape of a mortally ill Artaud screaming and beating on a drum. One rhetorical effect of the rise of "audio drama" has been to make all this so-called radio drama seem clearer than it really is and thereby generate an artificial rhetoric of novelty. Practitioners often pick moments of technological shift to "redefine" the medium along these lines: from point-to-point wireless to "broadcast" radio in the 1920s, when radio plays first emerged in the United States and abroad; from single-station radio to network radio in the 1930s, when experimental workshops arose; from radio to record in the 1970s, when stereophonic fantasy and science fiction took hold; and from analog to digital today.

These are also moments when critical traditions have suddenly emerged, in the form of radio columns in *Variety*, *The Nation*, the *Washington Post*, and the *New York Times* in the golden age and in the form of podcast brunch clubs, twitter feeds, and critical venues online in the 2010s. Often, it is the very lack of a sustained discourse that generated discourse. In 2017, for example, educator and producer Sarah Montague wrote a widely circulated essay advocating the development of radio criticism, arguing for its necessity for the work and for championing some of the things that other forms of art have—a pantheon of models, a separation between the creator and the creation, a sense of "aesthetic underpinnings" of a given broadcast. "Criticism is a hedge against invisibility," Montague argued. "It gives mass and intellectual weight to existing forms that help them to thrive."[35] The essay has had a significant influence on a field that was suddenly flush with cash but without an obvious way to make merits intelligible. In a 2019 keynote address at the Sound Education conference in Boston two years later, Pushkin Industries vice president Julia Barton quoted the essay, lamenting the lack of a critical idiom for podcast work such as the kind her company developed.

It is not really true that a critical idiom did not exist, only that it couldn't convey prestige or drive listeners in the way many practitioners wanted. Still, discourse developed rapidly in the late 2010s, with radio historians releasing many publications in these years, including the first books on podcasting that had "crossover" appeal beyond the academy (Dario Llinares, Neil Fox, and Richard Berry's *Podcasting: New Aural Cultures and Digital Media* in 2018, Lance Dann and Martin Spinelli's *Podcasting: The Audio Media Revolution* in 2019, Andrew Bottomley's *Sound Streams: A Cultural History of Radio-Internet Convergence* in 2020) and coverage of the genre in *RadioDoc Review* and Bello Collective. By the end of the decade, Galen Beebe pointed out that popular criticism on podcasting in general, and even podcast fiction specifically, was regularly appearing in the *New York Times*, *L.A. Times*, *The Appeal*, the *American Prospect*, the *Financial Times*, *Vice*, *Vulture*, the *Podcast Review*, and many more. An interesting aspect of this "lack of criticism" lament, however, is the fact that the points Montague made have actually been made several times before. There is an uncanny resemblance in tone, for instance, between Montague's 2017 essay for *Serendipity* and a similar essay by longtime *Variety* radio critic Bob Landry's called "The Improbability of Radio Criticism," published in *Hollywood Quarterly* way back in 1946. Like Montague, Landry found little appetite out there for a radio criticism worthy of the name, in part because of the sheer volume of radio that was being produced. He concluded the ideal critic would have to be bedridden to keep up. The two essays suggest that to lament radio's lack of a critical tradition is one way to join that critical tradition.[36]

This fallacy—the sense that radio history simply isn't there or isn't good enough to pay attention to—is not unique to the contemporary moment, nor was it limited to critics. On the contrary, practitioners have evinced the same attitude in the discourse of radio for the better part of a century. When asked by a *Washington Post* reporter about what to expect from experimental radio in 1936, a dozen years after radio dramas had begun airing all around the country, *Columbia Workshop* producer Irving Reis explained that the medium was entirely fresh territory. "We don't know what radio can do yet," he explained, promising to put the medium "through its paces" for the first time.[37] Four years later, in the foreword to one of the first radio directing textbooks written in the United States, director Earle McGill expressed his relief that finally, after eighteen years, radio directing was at last acquiring "dignity" for the very first time.[38] Nearly three-quarters of a century later, after countless radio plays had been presented around the world, producer Eli Horowitz of *Homecoming* said almost the exact same thing in an uncannily similar interview in the *New York Times*. "We're trying to explore what the form can be, while also trying to be conscious of what it is now," he said, also explaining to *Vanity Fair* during the same publicity push that drama podcasting was just at "the tippiest tip of the iceberg."[39] The author of the article went so far as to call this hundred-year-old genre the "new frontier" and a form that does not yet have a name. Just as Montague joins ranks with Landry in the ranks of critical amnesiacs, Horowitz and the other creators joined ranks with Reis. Both dramatists were in their contexts so focused on an apparent shift in material conditions of production (network radio for Reis, ease of digital distribution for Horowitz) that what was for both writers admittedly an "old" medium was re-mystified, largely out of a desire to promote the form in the popular press.

And this attitude is not exclusive to the "golden" and "modern" ages. Over the past half century, the only thing more reliable than the intermittent revival of radio drama is the insistence that what ensues is not really a revival but entirely new, as many creators have historically wished for their work to be heard as something other than "old-time radio." As Eleanor Patterson has shown, as early as 1963 (just a heartbeat after "old-time radio" had ended) radio producers were trying to bring back the form, while framing the project as something completely new. That year, producer Frank McGuire of ABC promised the new work would not "represent a return to old-time radio drama" while company president Robert Pauley recognized that "old radio drama didn't fit today's pace."[40] Compare these statements to those of Radiotopia head Julie Shapiro, who in 2016 told *Vanity Fair*, "We're starting to get away from the idea of the old-school radio drama with a capital R and a capital D" and instead "a more contemporary sense is developing of what audio fiction can be."[41] A nearly identical claim accompanies press reports of virtually every revival, from NPR's *Earplay* in 1972 and CBC's *Nightfall* in 1980 to the *Hollywood Theater of the Ear* in 1993 and on.

Even former boosters of the medium have historically tended to look backward uncharitably, some of them quite bitterly. In 1944, just seven years after creating *The Fall of the City*, the landmark literary play of the 1930s, poet Archibald MacLeish wrote that all efforts to produce radio dramas with the perspective needed to "give breath and presentness" to the spoken word had vanished.[42] In 1948, reflecting back on his nineteen years as head of production for radio plays at the BBC, dramatist Val Gielgud observed that in all that time he had failed to discover "a minimum of first-rate work" or to establish a "real school" of radio dramatists.[43] In a 2006 treatise on his many innovative radio techniques, *Hollywood Theater* director and producer Yuri Rasovsky wrote, "Unfortunately, while more audio drama is being produced now than in American radio's heyday, not much of it is worth listening to, much less keeping in a permanent collection."[44] Ironically, those who defend audio drama tend to be those who only dabble in it, for whom it is forever new, while no one practices the arts of amnesia more adeptly than those whose legacy rests on its remedy.

There are ramifications to this situation, in which the aesthetics of amnesia affects critics and creators alike as they conceptualize their medium. One result is that it tends to put creators and scholars at odds. In recent years, many in the scholarly community have been exploring the qualities of classical radio drama, recapturing the art of production, writing, direction, and acting and explaining how these qualities spoke to a variety of social groups. In academic circles, we are often working against outright dismissals of this material. It is true that old radio is *The Shadow* and *Ozzie and Harriet*, but there is a lot more to those shows than meets the ear, and, besides, radio drama has given us works that range from the enigmatic to the soulful (Beckett's "All That Fall" and Rasovsky's "The Dybbuk"), resonant pieces about the modern condition (Pare Lorentz's "Ecce Homo" and Stoppard's "Albert's Bridge"), and compelling atmospheres (From Orson Welles's "Hell on Ice" to Erik Bauersfeld's "Object Piece"). Yet it still conjures soft-focus images of unimaginative hacks with coconut halves poised to make the sound of horse hooves. Even works like *Dragnet*, *Suspense*, *Gunsmoke*, *Dimension X*, and *Calling All Cars*, which come across as overly mannered, have a depth revealed by dedicated and careful listening, a practice that is growing more and more common in academic settings.[45]

Curiously, the vogue of audio drama triggers the need for that very activity. Just as what we now think of as the masterpieces of film noir were deemed cheap "gangster pictures" until their aesthetics were rethought by French and American filmmakers a generation later, the new audio drama has a chance to redeem everything dismissed, parodied, and derided about traditional radio.[46] To take an example, the "Thistle Man," a haunting creature stalking the narrator of *Alice Isn't Dead* across the country, becomes more interesting when he is compared to his distant relative, the titular character in Lucille Fletcher's 1942 radio play "The Hitch-Hiker," who likewise chases the narrator across a ghostly nation, appearing over and over again by the side of the road. This is but one of hundreds of possible connections that could be made between the age of radio and the age of podcasting, to "cure" the amnesia by a process of critical anamnesis. Indeed, the disembodied voices from beyond the grave, brain experiments, fantasies of inanimate life, undead creatures, dystopias, time paradoxes, mass disappearances, alien invasions, love stories, and conspiracies that form the stuff of contemporary audio storytelling are not very far from the themes, techniques, and tropes that preoccupied radio in the past. And programming that was underwritten by ads for mattress delivery start-ups, meal kit services, and underwear industry "disruptors" in the 2010s is not very different from programming that sold beauty soap, boot black, and ironized yeast tablets generations before. The challenge is to see this coincidence as an opportunity, to transform the relationship between "old" radio drama and "new" audio drama from one of mutual antipathy to one of mutual vindication.

Anamnestic Criticism

Naturally, audio drama's many vehement disavowals of history make the prospect of historicizing it quite alluring. What would a criticism that does so look like, an approach to engaging with the audio drama podcasts of the late 2010s that shows their features pointing backward rather than forward, drawing out its latent poetic retrospection? Is there an available analytical stance that, instead of antagonizing its object by pedantically scolding bad memory, works in a restorative way, a critical anamnesis tasked with remembering on a podcast's behalf? To illustrate this idea, let me return to the three pieces linked to *Serendipity* discussed at the beginning of this chapter. While they do seem entirely new in character, a case can also be made that these audio dramas have interesting roots in traditional radio dramas. To begin with, "Tape Delay," "Movies," and "Every Heart" all used field recording rather than only studio work, joining a long tradition that bears comparison to the recent radio plays of John Dryden, such as *Pandemic* and *The Day Lehman Died*, and the pioneering experiments of Tom Lopez on his ZBS network in the 1970s (pieces like his *Travels with Jack* series brought binaural 3D audio recording to audio drama for the first time) and that might go all the way back to writer Norman Corwin's wire recording projects for the United Nations in the 1940s and earlier.[47] The emphasis on DAW-based stream of consciousness discussed above, moreover, is in some ways only a multiplication of techniques pioneered in the 1930s, when radio shifted from being a theater "in" the mind to become a theater "about" the mind, as I've argued, by foregrounding interiority, stream of consciousness, and psychology as many plays of the postwar period did.[48] The highly layered sound of plays like "Every Heart" and "Movies" would also have been at home in the 1980s among the experimental features of Marjorie Van Halteren, producer of the landmark 1985 *Breakdown and Back* series that attempted to depict mental illness sonically, or even in the 1970s, among experiments made by Erik Bauersfeld at KPFA in Berkeley, where the renowned auteur played often with layered and multiple voicing in his adaptations of Gogol's "Diary of a Madman" and Woolf's "A Haunted House" on his show *Black Mass*.

While audio fiction themes resonate with realized radio dramas, some of the formal and technical qualities that make "Tape Delay," "Movies," and "Every Heart" what they are also have roots in unrealized aspirations of earlier eras.[49] In the 1950s the BBC's legendary Donald McWhinnie, one of the most accomplished radio creators of his era, advocated for a radio that evokes rather than depicts consciousness; in praising legendary radio writer Giles Cooper, McWhinnie wrote of the latter's ability to create distilled experience "crystallized into a sound-complex; words, rhythms, evocative noises, fused into a kind of musical score which constantly stimulates the ear and the imagination." This is just what "Every Heart" aims to accomplish.[50] In a 2011 radio guidebook, critics Richard Hand and Mary Traynor identified the blend of "dramatic intent and musical technique," linking this impulse to the sort of radio drama pursued frequently in Germany and intermittently in the UK and calling for its expansion.[51] That is exactly what the concluding passage of "Movies" offers when the protagonist is ghosted by her lover and driven manic, a section depicted through a series of layered memories and desires overtop a heartbeat and repeated text messages by the narrator to her lover, the time ticking by, phone messages without answer piling up, until desire and frustration become a rhythm, a telephone ring gets suspended in time, and a bass line enters the score to set it into a maddening groove. The fact that these shows are anthologies matters, too. *Serendipity* and *The Truth* could with some justice have presented themselves as successors to venerated playhouses of the previous century—*Studio One* in the 1940s, the *BBC Radiophonic Workshop* in the 1950s, *Earplay* in the 1970s—that similarly sought highbrow experimental work and collectively bent the aesthetic trajectory of radio by framing what counts as mature, edgy, and medium specific. In his 1940 book on radio directing, the veteran dramatist Earle McGill wrote that workshop radio of the 1930s was important because without it "there should be no place to try out new forms, new uses of sound, new approaches to the problem of fitting words and music to the medium."[52] *Serendipity* and *The Truth* wanted to do that, too, compensating for excesses of genre serials while foregrounding literary formalism, in exactly the same way as their predecessors.

There is also a notable concept these plays share that is curiously old-fashioned. Just as "Tape Delay," "Movies," and "Every Heart" have the same point of departure, with a couple meeting in New York, they also conclude in the same way, with passages that explore telephone speech and miscommunication, along with alienation and paranoia about technologically mediated voices. That theme connects to a well-known 1997 piece by

Joe Frank, "Phone Therapy," in which the experimental broadcaster used fiction and fact to explore psychotherapeutic group phone calls that walk the line between fiction and fact.[53] The centralization of the telephone also resembles the most famous radio thriller of all time, Lucille Fletcher's 1943 *Suspense* play "Sorry, Wrong Number," in which Mrs. Stevenson, an invalid played by Agnes Moorehead, overhears a murder planned on a crossed line and then makes a series of phone calls to annoyed authorities to try to stop it; Fletcher intended the play to take the "perspective" of a telephone. You could say the same thing about the concluding moments of "Tape Delay," "Movies," and "Every Heart," plays as much "about" the mysterious thinness of telephonic miscommunication as they are "about" the superabundance of DAW-based automation software. Indeed, these pieces may represent not the beginning of a new cycle of audio fiction but the end of a history of the microgenre of the "telephone play," one long bound up with issues of gender, frustrated desire, and the fear of death. Just as the relatively new prevalence of analog telephony had evoked these thoughts for Fletcher, the relative new prevalence of smartphone telephony brought it forward for contemporary dramatists. Viewed in this way, these pieces suggest the persistence of a cluster of narrative features echoing down the tradition rather than a new form.

The purpose of this exercise has not been merely to say that everything seemingly new has been done before but rather to suggest that one job of the critic is to think alongside these works in anamnestic ways that vivify and also complicate their amnesiac tendencies, to speak in a certain sense *for* the archive that is so deliberately set aesthetically external to the work.[54] Indeed, the excessive way in which so many audio dramatists frame their intervention in the field as novel suggests a critical countermove into the past through unconscious latencies and remainders, showing the acts of forgetting that make possible acts of creation. To make radio plays "for the twenty-first century" is, after all, already to have some specter in your mind about plays "of the twentieth century," likely derived from vaguely recalled archival examples that draw an external perimeter. In this way, the aesthetics of amnesia both excludes and points at the archive at once, like the cliché amnesiac remaking themself as something new, yet all the while obsessed with the previous self from whom they have supposedly decided to start over. The audio critic who undertakes anamnestic reminding represents a fulfillment of a secret desire within the aesthetics of amnesia.

So much for anamnesis when it comes to high-end "non-non-fiction." Let me shift now to some key independent productions from beyond the

Third Coast orbit, looking at a few features that will help situate these works in longer arcs of audio drama history as well. In indie audio fiction, a common strength I have always enjoyed is the sheer complexity of its set pieces, ranging from teenagers whose heads open up as if on hinges and get lost in portals under an old New England house (*Locke & Key*) to genetically engineered giants tearing a Mumbai slum apart in pursuit of a messianic immortal (*The Leviathan Chronicles*) and ancient stairways in hidden ice caves made of thousands of human teeth (*The White Vault*). While most prominent in horror plays, this sense of complexity was also shared with more comic and speculative fiction serials—*Our Fair City*, *Greater Boston*, *Wolf 359*, *Uncle Bertie's Botanarium*, and others—that enjoy stretching out into richly imagined settings. Considering their scant resources, these plays are surprisingly big, fast, and pleasurable. Sonically, all of them resemble Douglas Adams's landmark 1978 *Hitchhiker's Guide to the Galaxy* in the sense that, like Adams's plays, these pieces are adept at world building.[55] And like his works, they are also philosophical, fascinated with scale, time, and space.

A common difficulty among these programs has to do with the way they incorporate violence in highly articulated digital audio. In the classical radio period, this was less difficult—there is a smile inside every scream in plays by Arch Oboler on *Lights Out!* and Wyllis Cooper on *Quiet Please*, whose most terrifying broadcasts back in the 1940s were also their most fun. That is not the case in more recent work, as leading programs tend to either underplay or overplay horror. There is a rape and murder scene in the first five minutes of *Locke & Key*, something the whimsical music and sound design do not prepare us for. In the sixth episode of *Bronzeville*, Lawrence Fishburne's Curtis Randolph, a community patriarch driven mad by the murder of his wife, takes a razor to a suspicious hood who may have information, but the sequence happens well off-mic, diminishing its dramatic power and capacity to bear moral complication. In the first episode of *Inside Psycho*, meanwhile, the sound designers go very far into explicitly depicting the murders by Ed Gein, including the ghoulish sound of the body of victim Mary Hogan being ripped apart, a veritable sequence of torture porn. Because audio had been governed by network and FCC rules when it was mostly on air, and since other norms focus on visual depiction, many creators had a difficult time knowing what counts as too "graphic" a depiction for a given audience and how to prepare listeners for it. Over time, many podcasts in the independent space took it upon themselves to issue content warnings in an act of generosity (the first I remember was *Ars Paradoxica*), perhaps because indie productions were so close to their fans (many producers made work for one another) that these communities evolved an ethic of mutual care that commercially driven shows would eventually follow rather than lead. *Someone Dies in This Elevator* used these warnings in an ingeniously creative manner, at once warning listeners about content (e.g., "This episode contains self-harm"), signaling the humane and progressive politics of the show ("contains structural ablism"), as well as keeping the premise of the series going as a meta-gag ("contains death in an elevator").

The interest in complex settings and the unsure relation to violence are two sides of the same coin, since both engage with the problem of adequate depiction as such, with these pieces showing a deep awareness of the variability of the existence of objects, events, and settings in the worlds they create; many pieces are exciting because the things they depict are so *there*, while others feel off because the things they depict are too *there*. In an essential 1945 treatise on the poetics of radio, radio writer and historian Erik Barnouw wrote about this very phenomenon, what he called the "shifting mind-world" of radio drama, in which scenery, costumes, smells, heat, and even characters may be vivid one moment, fade another, and then snap back into mental existence.[56] For Barnouw, while many media rely on sharp differentiation between that which exists and which does not (what is onstage or not, on the page or not, on the screen or not), radio drama is much more ontologically supple, because it is just as easy for listeners to take on board uncertainties as it is to subscribe to anything with "full existence" in the space of the fiction. By dint of radio's nonvisuality and intangibility, as well as the writer's sense of how to invite listeners to project their own ideas (or to withhold that same capacity through the instruction of dialogue, effects, and music), radio drama often features the "semi-existence" and "potential existence" of objects and characters in dramatic space.[57] To create for audio is to write sentences that always have open ends, to draw pictures whose frames are only there sometimes. We don't hold objects and events to the same level of predictable materiality as we do in other media; we sometimes even forget some of the people we're listening to are there. In audio, in short, there is no *mise-en-scène* in the traditional meaning. Instead, what we think of as *mise-en-scène* obeys the logic of contingency, of hint, and of hallucination. Scholar Mary Louise Hill explains this effect by borrowing a term from translation studies, explaining that as we listen and try on a series of unfixed meanings in audio, it's less like encountering *mise-en-scène* and more like encountering *mise-en-jeu*.[58]

Bearing that in mind opens up dynamics in many dramatic podcasts. In Lauren Shippen's *The Bright Sessions*, for instance, psychiatrist Dr. Joan Bright specializes in working with "atypical" patients who have superhuman powers as time travelers or empaths. In yet another podcast that begins by taking place in tapes, each session ends with hints about a subtext—"the project" for which various patients are "assets"—anticipating that much of what the listener believes to be the real story "happens" in their own conjectures before it opens up in later seasons. Critical praise of the show, while centering on emotional depth, also often notes a sense of scale behind that psychological narrative, with its "sprawling universe" and "blockbuster" plot twists, something we are invited to construct incrementally.[59] In Justin McLachlan's *EOS 10*, a comedy about doctors in a space station, one sequence features protagonist Dr. Dalias poisoned by an alien lover with a xenopharmacological aphrodisiac that gives him a priapism. Much of the two-episode story arc is taken up by scenes of embarrassment with the stubborn erection, something made funnier by the absence of a visual (to a colleague: "My eyes are up here, Doctor"). Importantly, we can from time to time forget about the existence of his discomfort as they seek a cure. The joke works because something that exists occasionally lapses into "semi-existence" and then returns to existence, which would be hard to accomplish in visual media that seldom lets us forget what we see.

A fascinating game of presence and absence is also played in *Wolf 359*, a beloved science fiction comedy-thriller by Gabriel Urbina and Sarah Shachat that takes place in a distant outpost around an alien star. In the twentieth episode there is a long scene in which four mutually suspicious characters spin elaborate plot theories accusing one another of stealing a missing screwdriver, something whose presence or absence audio cannot "objectively" confirm, until we are told at last that actually it was swiped by a plant monster hiding in the ventilation—a creature we hadn't heard about for fifteen episodes and was thus forgotten. The scene solves a present absence by using another absence it has cultivated in its distracted listener, a brilliant long con and a unique way of solving the traditional locked room mystery construction. A more recent series that features a sophisticated understanding of concreteness and ambiguity is *Unwell*, Jeffrey Gardner and Eleanor Hyde's intimate social exploration of a mysterious Ohio town. In illustrating its rooming houses, observatory, forests and other spaces, the sound design makes scrupulous use of stereo spreads, vocal muffling of characters talking while separated by doors or down corridors, as well as meticulous floor creaks, wind chimes and radiator sounds.

By the time ghostly voices, trapped echoes, disappearing doors and other uncanny events begin to occur, our subconscious sense of the town is photorealistic in style. In this way, the show paints its naturalism and its supernaturalism in alternating strokes; before the long the setting isn't just "a character" according to the old cliché but a plasm that makes its characters and their sense of their own elastic reality gel. Finally, take Daniel Manning and Mischa Stanton's *Ars Paradoxica*, a historical drama in which scientist Sally Grissom is transported back to the 1940s and a government science initiative, where she begins to create a "timepiece" that sets off an alternate timeline employed by a shadowy government agency. The series asks us to imagine things that are either difficult, such as the plots of the agency made up of time travelers who get their orders from bosses in variable futures, or nearly impossible, such as a bullet that travels back through time to strike the same man in two different timelines: the widow of a shooting victim confronts Grissom about the death of her husband and then shoots at her but misses and sends a bullet through the time device, thereby killing the husband in the past whose killing she had come to avenge in the present.

These creators intuit that the seeming impoverishment of sound as a medium is its greatest deception; in fact, the genre has several advantages in the dynamics of presence and absence that stage directors, TV producers, and filmmakers cannot employ. If the problem is how to depict happenings in audio that most listeners are used to experiencing with images, then the discovery is that audio provides more leeway, not less. "The difficulty of scene-changing in broadcasting," wrote Rudolf Arnheim, one of radio's first theorists, "consists paradoxically of its being too easy."[60] For Arnheim, the mistaken impression that something (i.e., visuals) is "missing" in radio masked the deeper strength that because audio offered unity to the medium, "missingness" itself could be used as an effect of depiction rather than as an ontological state. In how they solve the problem of the ease of depiction in narration, then, these programs share a capacity to leverage the property of partial existence for ends that range from exposition to affective response, to use the non-depicted and semi-depicted intrepidly, which is the reason why complex depictions were so startlingly easy, and violent ones so hard to measure correctly. In this way, they are dramas of ambiguous presence and, as such, capture something of the frenzy around the tangible in contemporary media practices. I am not sure if these dramatists had any familiarity with Barnouw or Arnheim, but that is beside the point. These ideas—written as "lessons learned" in radio drama of the 1930s and 1940s—work well with this eminently contemporary material, as if the experiences earned by dramatists generations ago are still somehow around in the muscle memory of creative audio. For every audio drama that starts from scratch and feels as if it suffers from amnesia, an anamnestic opportunity appears.

In fact, like the *Ars Paradoxica* bullet, engaging audio drama with radio drama allows us to move both forward and backward in time, connecting contexts and traditions. Rather than hearing the way the work of Dirk Maggs has influenced the sound of such adventure programs as *We're Alive* and *Locke & Key*, this approach would consider what makes Maggs's style so durable, how it relates to the techniques of *I Love a Mystery* in the 1940s and *The Fourth Tower of Inverness* in the 1970s, two pieces that likewise solve complex action problems in closed spatial environments. When X, the unnamed character-narrator of Issa Rae's *Fruit*, spends whole episodes playing games of cat and mouse to avoid verbally revealing the secret of his sexuality while relentlessly telling us about it, the dramatists are employing a structure common in "undercover man" thrillers, from *I Was a Communist for the FBI* and *Counterspy* in the 1950s to Tom Stoppard's *The Dog It Was That Died* in the 1980s and Robert Forrest's adaptation of John le Carré's *The Spy Who Came In From the Cold* in 2009. Heard in the context of *Fruit*, it becomes clear how White cis heteronormativity was a subtext of *I Was a Communist* in the 1950s; heard alongside *I Was a Communist*, politics becomes a clearer theme in *Fruit*. Consider *Bronzeville*. In offering copious vantage points on a highly specific place and time, the series employs what I've called the kaleidosonic style, one in which we hear rapid shifts between a number of shallow scenes across a social geography, drawing in the voices of fortune tellers and preachers, hustlers and cops, bankers and gangsters, new arrivals to Chicago from the Great Migration, and others. Through segues and an emphasis on the present tense, characters rhetorically form a bonded group, as the rhetoric of a kaleidosonic program is one in which voices accumulate into a coalition, emphasizing equality and pluralism, a style perfected by Norman Corwin in the mid-twentieth century. *Bronzeville*'s focus on Black voices, however, also gives us a new way to listen to Corwin, whose work (with the exception of his 1939 *Ballad for Americans* with Paul Robeson) largely excluded Black voices from its monumental national tapestries. A dramaturgy that loses its amnesia can not only redeem predecessors but also deepen critique of those predecessors.

To conclude on this point, I would like to return to three "watershed" podcasts of the period: *The Black Tapes*, *The Message*, and *Limetown*, plays notable on the surface for adopting the female-investigative-podcaster-who-is-not-who-she-appears-to-be, perhaps the most common protagonist convention among serials of the 2010s. Beneath that surface, however, a lot is going on. All three use sounds that walk the boundary between illustration and affect, taking full advantage of the "semi-existence" that audiophonic narratives allow. Take *Limetown*. Most episodes of the program follow a pattern. In each episode, reporter Lia Haddock makes contact with a survivor of the Limetown disappearance who has only vague, semi-wiped memories from that time. Eventually the survivor gives a long monologue about their time in the mysterious town but generally refuses to answer real questions. At no point are there indications that the play segues back in time to the scene of the Limetown disappearance, as audioposition remains in whatever setting the survivor is speaking to us in the present (secret waterfalls, trailers, hidden lairs, closed up shops, foreign parks).

But something odd happens in the audio. During each monologue, a sonic element of the story-within-the-story is reproduced in the underscore with just enough stylization to make it seem "there" but also obviously nonpresent. A Limetown cleaner remembers the sound of pages being torn from twin drawing easels during a mind-melding experiment, and we hear it; a man who raises pigs for experiments remembers the sound of an alarm bell that we catch in the soundscape; a survivor recalls the sound of a mob leader with a megaphone as a funeral pyre is lit during the collapse of the town. It is a sonic code for memory, the auditory equivalent of a thought bubble, and it gives the scene a new tier for its depiction, a mnemonic sonic mezzanine. Theorist Marie-Laure Ryan has an approach to narrative structure that is useful for this, noting we should analyze narratives in a way that shows both how they "frame" various sets of events, using borders, and also how they "stack" sequences, like trays over one another. In this case, Haddock's narrative creates a frame, as when she meets her various interlocutors and their narration creates an illocutionary border as the interlocutor begins narrating an embedded narrative. Meanwhile, the sound design stacks a level of depicted action from the past on top of the narration, ontologically separated from the others as evidenced by the fact that the narrating Limetown survivor does not seem themself inside the space of depiction as they narrate. In this way, the deployment of a diegetic sound form transforms a moment of framing into a moment of stacking.[61]

This use of sound resonates with the titular "Message" in *The Message*, an abstract sound composition that becomes a dramatic object in the play, taking on a life of its own as the codecracking team tries to decide what to do with it. The composition is the centerpiece of the series, standing within but also apart. There is a similar recording in *The Black Tapes*, one of whose storylines involves a cursed composition (the "Un-sound") that supposedly brings death once heard. Where the stylized sound events in *Limetown* break the naturalistic sonic language of the program, creating an abstract sense of the past, "The Message" and "The Un-sound" break the frame of the narrative by proposing they could be harmful to the listener herself, too. Elsewhere I've made the argument that during World War II, classical radio drama embraced sounds that work as "signals" in that they urge and compel behaviors; these sounds do not merely illustrate some exterior physical event but are presented as if we are supine before their awesome influence, which alters or distorts the field of the "reality" of the fiction in which they resound.[62] In all three of these programs, sounds work in a similar way, like viruses between consciousnesses, and between the inside and the outside of the fiction, giving these plays a psychological dimension. Each sound has a quality of *want* in its relation to us. It is a remarkable mutation of the dramatic preoccupations of the past and indeed a late metaphor for the fantasy of radio itself, a medium that has done more to change the fabric of reality through sound than any other, to cast spells, to alter minds, to connect bodies.

Coming Home

In the previous two sections I have been exploring what a critical anamnesis might achieve, unlocking some of the latent material in contemporary work, while at the same time allowing for a rethinking of how classical audio itself might be reapproached. To complete this thought about audio drama "coming home" to radio, an essential series is, naturally, *Homecoming*. The 2016 season of the show deals with a soldier suffering from PTSD named Walter Cruz (Oscar Isaac), who is subjected to unethical testing of memory removal, and with his caseworker, Heidi Bergman (Catherine Keener), who first participates in the experiment and then develops feelings for Walter and undergoes a memory wipe once she comes to realize the damage the experiment has done. We meet Heidi both in the main series of events from the past and also many years after the fact, in entwined sequences. Thanks to this double timeline, over the course of the episodes, we hear through flashbacks of the gradual diminution of Walter's memory in the past sequence and in the present of the gradual restoration of Heidi's. While the show strongly echoes several *Suspense* plays of the 1940s about amnesia ("The Black Curtain" with Cary Grant; "Mission

Accomplished" with James Stewart; "The Search for Henri Lefebvre" with Orson Welles), it is particularly interesting in how it charts a double movement of past forgetting and present remembering, an aesthetics of amnesia working in both directions.

The podcast's most remarkable classical formal element is that it rests so much of its sound and narrative design on what I call audioposition (where we are according to what we hear) as a structural device. As I explained in chapter 2, attending to audioposition in radio and audio work is so important because, after all, there is no "natural" where-to-listen-from in the world of these fictions. Unlike the situation with audiovisual media, in pure audio we can't ask what events "should" sound like according to the position of the camera, proscenium, or other correlated element forming a window onto the scene that could serve as the basis for conceptualizing a "realistic" miking approach. Every microphone opens a world uncontradicted by some other readily extracted element, until it is complicated by a second microphone or some other production layering. A space may not sound acoustically similar to what the dialogue suggests it is, but both sound and dialogue pass through the same sensory aperture, so even if it is unrealistic, that does not affect our perspective. This is why the rhetoric of position matters deeply in radio and audio drama formats, since it is not defined by anything else and touches every aspect of the craft of audio, from writing to soundscape, from Foley to vocals. In interpreting traditional radio, audioposition helps to frame moral alignments, makes it possible to feel suspense, to teleport from place to place, to convey diegetic time, and produce any number of other effects. Without a sense of where, there can be no "identification" with one character over another, no sense of when, what, or why, let alone a sense of uh-oh, whew, or yuck.

With that in mind, let us explore the three audiopositions that we hear most often in *Homecoming*. In the first audioposition scheme, we hear interview tapes recorded at the military base near Tampa, Florida, in which Heidi Bergman gets to know Walter Cruz, who is there to learn to deal with his "bad thoughts" after the loss of a member of his unit who died horribly in a senseless incident at a traffic control point. The dialogue between Heidi and Walter is warm and flirtatious, intended to chronicle the creation of rapport. In an early episode, Walter tells Heidi of a prank in which he conned a soldier in his unit into believing in a made-up sequel to the film *Titanic*, before mentioning a comrade's death in an understated way. As the counseling sessions continue, Heidi helps Cruz vocalize his memory, express incredulity toward the meaning of his service at the checkpoint, and plan for a future in a new career. The office space in which all this occurs is three-dimensional, carved out by prominent sonic elements, including a backdrop of the bubbling of a fish tank, as well as the exterior call of a bird (a "protected species" we are told) and the interface of the tape recorder itself, which is often audibly handled in a way that produces a palpable wall between the space of the recorded event and that of its reception. The room is coded as domestic in Heidi and Walter's interplay, but it is also increasingly carceral: surveillance mics are everywhere in the Tampa facility, there is a false "escape" sequence in which Cruz and his roommate run away only to discover themselves in a retirement community, and Heidi and Walter daydream a road trip, the sort imagined by prisoners. The facility is a kind of traffic control point for memory, a River Lethe.

In the second repeated setting, we hear a series of internet-based phone calls between Heidi and her boss, Colin Belfast (David Schwimmer), who is traveling the globe as he raises funds and political support for their project, the "Homecoming Initiative." Thanks to exposition in these calls, we gradually learn that Heidi's emotional labor with Cruz is only a ruse; the true goal of the project isn't to rehabilitate soldiers coming home to society but to entirely erase traumatic memory through a combination of drugs, administered through food in the mess hall, that will enable them to return to the battlefield more quickly. Like *Limetown*, *The Message*, and *The Black Tapes*, *Homecoming* meditates on a classically radiophonic question, mediated mind control. In contrast with the first setting, this one is overtly hostile. Colin takes Heidi to task for minor rebellions among Walter's peers, accusing her of "shitting on their results" and "erratic emotional outbursts." Our audioposition is in the 2D claustrophobia of telephonic space, where Heidi can't get a word in edgewise thanks to a sonic flatness emphasized by Colin's allusions to deeper spaces (the Detroit airport, the roads of Dubai) that we can't quite hear, by his constant technical problems, and by his penchant for using these conversations to play surveillance tapes from the Tampa facility that add to a growing paranoia. This is an allusion to "Sorry, Wrong Number," with Heidi's voice thwarted in a flat vocal space in the same way as invalid Mrs. Stevenson is ritually dismissed by gaslighting authorities in her pleas for help in Lucille Fletcher's classic play. Note that in both of *Homecoming*'s first two settings our audioposition is *within* a technological device—we are "inside" a recorder in the office and then "inside" Heidi's laptop making a call—but in the former we lose a sense of that mediation and come to feel intimate companionship with nearby interlocutors. The strongest sense of intimacy is achieved not through the absence of mediator altogether but by its dissolution over the course of a scene. Meanwhile, in the phone call scenes in which a similarly prominent mediator refuses to vanish, there is an accompanying sense of a highly gendered asymmetry of power. The concluding episode of the first season has a nice reversal of this effect, when Colin grovels for help to his own superior, Audrey Temple (Amy Sedaris), and receives only a comeuppance.

So far we have two time-spaces with their own unique signatures: a series of tapes recorded by a handheld device in a therapist's office in the past and a series of phone conversations recorded in between those sessions through a computer, also in the past. In their use of archival audio and slipshod recording technique, as well as the exorbitant way in which the recording mechanism is alluded to in dialogue, both settings are highly legible as "podcast" in their style. The third setting is different. It is characterized by a time period rather than by a place. Moreover, there is no flimsy narrative pretext to account for the existence of the recording we are listening to by attaching it to a diegetic recording device. Instead, we are in the scene, as if following intimately over the shoulder of Heidi and segueing across the space of the fiction alongside her.

It is five years later, and Heidi is working at a diner near her elderly mother, whom she regularly visits at a nursing home. Thomas (David Cross), an agent of the Department of Defense, arrives at the diner and begins probing her memories of the Homecoming Initiative, which are vague and deeply buried. Over several episodes, we begin to understand that her own memories had been wiped by drug therapy, as Colin (newly anonymous to her) comes back into her life (another homecoming) and she returns to the scene of the Homecoming Initiative experiments. At first Heidi has trouble remembering, but the call of the bird outside her office brings parts of it back. This setting sequence sounds most like that of a classical radio drama compared to the other two. It also teaches you how to listen to dramas of that period, by emphasizing depths (Heidi moving around the restaurant, through the nursing home, and in a parking lot and restaurant; a long silence in Thomas's car), as well as spaces of dynamic motion, such as a Ferris wheel scene and Heidi's escape from Colin across passing traffic. In the present we are attached to a body, not just adjacent to a voice in conversation, and the world that surrounds that body contains fragmented, refracted evidence of the story. The amusement park ride is a (somewhat forced) metaphor for Heidi's predicament, caught by circular forces. At the nursing home, we repeatedly hear a TV with a news story

Figure 13. A spectrogram of episode 5 of *Homecoming*, with the waveform minimized. I have identified and labeled 8 separate sections. Decoded in Izotope RX6. Screen capture by the author.

about therapies to improve "brain fitness" deep in the background (you need good headphones to hear it), a hint at the speculative drug that is the premise of the action. In this way, story information is hidden away in the depths of the audio and the shapes of movement, forcing the listener to a level of auditory projection that nondramatic radio never requires.

To sum up, *Homecoming*'s three main settings are, respectively, three-dimensional (the office), two-dimensional (the phone), and four-dimensional (the present), a set of audioposition schemes that are constantly "talking" to us through and behind the events that take place across them. The first two sound like other audio dramas, while the third sounds like traditional radio drama. Thanks to the way space is managed by expansion and contraction, the series feels like it squeezes you and releases you over and over again. You can see these distinctions between scene types in a spectrogram view of the fifth episode in the series. The first segment (0:00 to about 7:30) in figure 13 takes place in Heidi's office, a full scene filled up with the sound of the broadband bubbling fishbowl and low end mic handling noise. The second (7:30 to 13:45) and fourth (16:00 to 18:00) segments show the present sequences, in scenes in which Belfast and Heidi ride a Ferris wheel and then head back toward Tampa together in a car. In both cases, you can see the punctuation of sonic events—dialogue, rattling carnival sounds, rumbling cars—that give the scene an openness and variability that indicates movement through space. The third (13:45 to 16:00) and fifth (18:00 to 21:30) segments, by contrast, are the two-dimensional world of the calls between Belfast and Heidi. That sounds thin, claustrophobic, limited to vocal speech. You can see the signal is compressed to fit between 100 and 5,000 hertz. The sixth sequence returns us to the present, with Heidi realizing suddenly that Belfast had led her back to the Tampa facility, a bird sound jogging her memory, which ends the episode on a cliffhanger (the last two sections are music and a post-show interview).

My point in this analysis is to emphasize that like many classic radio plays—Archibald MacLeish's *The Fall of the City*, Richard Hughes's *Comedy of Danger*, William Robson's "Three Skeleton Key"—the "what" of the drama only exists in the "where." In *Homecoming*, two-dimensional conversations are claustrophobic and choked (Heidi can hardly get a full sentence out) and four-dimensional life is precarious, full of as many gaps and spaces as her lost memory (she often speaks of alienation: "It's like it wasn't even me"), but the therapy room avoids both. That is, until Walter's mental conditioning has taken hold and he forgets the anecdote about *Titanic* (a sign of disaster) and the room becomes uncanny. The only comfortable space to occupy in the entire fiction has only had a false sense of safety all along; the atmosphere of a home has masked the reality of a prison, not vice versa. Walter can no longer remember the details of his trauma or much else. Most importantly, he no longer feels frustrated by the meaninglessness of his task overseas, and he becomes eager to reenlist.

Through its various devices, then—vanishing mediator, use of sonic depth to hide story information, and the variation in the rhetoric of dimension—the sound design manages exposition and mood, coaxing these two ordinarily separate dramatic elements into conspiracy. This is what makes this play so classical in spirit if not in content. *Homecoming* is using sound as rhetoric, and so it must necessarily teach you how to listen to it. In this way *Homecoming* set an impressive standard. In the coming years, a series like *Wolverine: The Long Night* would do something similar: concoct a sonic rhetoric unique to the narrative. Recorded with an ambisonic mic (which captures sound in a tetrahedral pattern of cardioid mics that allows for excellent postproduction manipulation of the signal), *Wolverine* uses a system in which large, beautifully recorded bars, forests, and ships in a remote Alaskan fishing town are set out for the listener, until the investigators searching for the eponymous hero come in and begin asking questions. As the interviewee begins to remember, the scene slowly draws down to a pinprick hole, and the speaker is put into a deeply shallow spotlight in which all the suppleness of her or his voice is intensified and drawn close. Like *Homecoming*, *Wolverine* uses mic technique to create a style, one that makes it legible to itself and has the concomitant benefit of instructing the listener, too.

In the end of the first season of *Homecoming*, Heidi realizes the drug program wasn't eliminating memories in a targeted manner but was brainwashing Walter and the others of unrelated memories. The "smart bomb" approach to only targeting traumatic memory had been a lie all along. Heidi decides to give Walter an extra portion of the experimental drug to make his return to the battlefield impossible and also to take it herself as a penance for her sins and to deny Colin success. We now know why the narrative gap between settings occurred. Heidi's homecoming to the Tampa office building five years later is the beginning of re-memory about all these events. The moment she hears the signal of the bird's call, something we had first noticed in the very first episode, she begins to remember in a moment of sudden anamnesis, and Thomas supplies her with the very tapes we have been listening to all along to jog her memory, thereby aligning her ear with ours in space and time; in this closing of the narrative loop, we know that Heidi could be a hidden simultaneous auditor, just like us. It is precisely by listening to her own "audio drama" that Heidi's "radio drama" begins to make sense, and as we listen for her to remember, we are remembering how to listen.

Pillow, Talk

By the fall of 2018, roughly three years after "The Revolution Will Not Be Televised" session with Ann Heppermann and Martin Johnson cited at the outset of this chapter, the world of podcast fiction was in full bloom, with serials that deepened many of the themes discussed above. Found footage stories like *Video Palace* and *Archive 81* were replaced by fragmented narratives, with podcasts emerging in quite different idioms on this theme—*Rose Drive*, a revenge story about a group of high school friends that updates the *Limetown* serial interview model; *The White Vault*, about a rescue mission gone missing in a frozen Scandinavian mine that introduces an external narrator to piece together evidence; and *Janus Descending*, about a xenoarchaeological mission presented to us by two narrators' logs issued to the audience out of chronological order. Shows focusing on Black history also soon emerged, including *Harlem Queen*, another show focused on northern African American enclaves during the Great Migration, an ambitious series by Yhane Washington Smith about 1920s numbers racket boss Stephanie

St. Clair, with appearances by actors playing Langston Hughes, Zora Neale Hurston, and Lucky Luciano, over a Bessie Smith soundtrack. Anthologies continued, including *The Amelia Project*, featuring the comic interactions of a series of jilted wives, exasperated reality stars, and vengeful robots keen to wipe away their past lives and gain new identities. Dream team shows emerged, too: Lauren Shippen from *The Bright Sessions*, Paul Bae from *The Black Tapes*, and Mischa Stanton from *Ars Paradoxica* began working on *Marvels*, while John Dryden of *The Message* and Shippen teamed up with Mark Henry Phillips, the sound designer for *Homecoming*, to produce *The Passenger List* for Radiotopia. *Unwell*, a gothic serial podcast by Jeffrey Gardner and Eleanor Hyde that explored the meaning of family, identity, and recovery in a strange small town in Ohio, won the BBC's first Podcast Drama award. Memory loss was a key theme; its first episode was titled "Homecoming."

Jonathan Mitchell's *The Truth*, meanwhile, released an ambitious multi-episode #MeToo-themed story about a disgraced TV host called *The Off Season*, while also gearing up for a Raymond Chandleresque detective story starring a Hollywood personal trainer, *The Body Genius*. Gimlet followed the success of *Homecoming* with *The Horror of Dolores Roach*, a Sweeney Todd–style story of a recent parolee returning to her gentrifying neighborhood starring Daphne Rubin-Vega and Bobby Cannavale; Roach gives massages to her clientele that end in murder, after which her lover-accomplice bakes the victims into his best-selling empanadas. Gimlet's *Sandra*, meanwhile, follows a lowly information worker named Helen (Alia Shawkat) as she takes on a role as a virtual assistant (Kristen Wiig), speaking to clients through the wall of an interface. And by Third Coast that year there were already previews out for a TV series based on *Homecoming*, as well as announcements about a similar *Limetown* deal to go on Facebook Live. In the coming years, more and more creators would be selling options for their podcasts to TV production companies. The revolution, it seemed, might be televised after all.

But none of these works earned a showcase at Third Coast that year. Instead, the keynote came from someone else, whose work built on the "audio fiction" *Serendipity* style with which this chapter began and that remained the sole vector by which audio drama found its way to the highest echelons of industry esteem: Kaitlin Prest's *The Shadows*, which she produced with Phoebe Wang for CBC podcasts. First conceived as an expanded revision of "Movies in Your Head," *The Shadows* is a six-part semiautobiographical work that took a year to research and create in and around Toronto. Prest spent eighty hours recording with one of the principal players alone. Set in the fictional city of Mont Yuron (an amalgam of Toronto, Montréal, and New York), the piece follows a queer puppeteer—named "Kaitlin Prest," played by herself—as she falls in and out of love with another puppeteer (Charlie, played by Mitchell Akiyama), while pining for a secret side lover (Devon, played by Johnny Spence) and featuring an imaginative passage on the erotic life of Devon's borrowed sweater, which comes to life with Wang's voice speaking directly to us.

The Shadows is less about love than it is about lovers as mutually problematic objects, evoking Lauren Berlant's notion of "cruel optimism." For Berlant, optimism begins with a desire or attachment in which the object, scene, or person seems to open a cluster of promises for us; optimism becomes cruel when even though the loss of promises can't be endured, the object that bears them threatens simultaneously our flourishing and even our well-being. The result of losing that object isn't just the loss of love but the loss of hope. "Where cruel optimism operates," she writes, "the very vitalizing or animating potency of an object/scene of desire contributes to the attrition of the very thriving that is supposed to be made possible in the work of attachment in the first place."[63] In its sound and approach to this theme, *The Shadows* is quite different from almost everything else in this chapter, and its aesthetics are, too. If shows like *Homecoming* represented one possible future for audio drama—as an amnesiac form about amnesiac characters—*The Shadows* seemed like an off-ramp that might have led to something else, an alternative vision for what the future of audio might be like, a queer feminist appropriation of the aesthetics of amnesia that mold it around nostalgia, embodiment, and objecthood. To conclude this chapter, I dig into this alternative aesthetic, showing how it swerves away from the aesthetics of amnesia while at the same time transforming it into something else.

That begins with its relation to heteronormative time. At the core of heteronormativity, as theorists such as Jack Halberstam have argued, is a timeline of life that is organized around family, longevity, security, and inheritance, from which any deviation is pathologized.[64] Overcoming that is a quandary in *The Shadows*; more than love, it is an accommodation with reproductive, normative time that Kaitlin struggles to rationalize to herself in a queer framework. This thematic struggle affects the podcast structurally. Across the piece we hear a number of shifts of audioposition from episode to episode. We "hear" episode 1 from Kaitlin's audioposition; episode 2 shows us the same events from Charlie's; episode 3 is heard from the audioposition of Sweater and overlaps with four separate episodes. To help us decode analeptic shifts in the disjointed narrative, repeated phrases act like red threads ("I want to say I love you, but I feel like it's too soon"; "Does that mean I'm your girlfriend?"), each appearing in a single chronology twice. In narratology, the time-shifting that Prest employs is called "internal retroversion." Things happen in the story, and we hear them from one perspective, and then in a later episode we circle back and hear them happen again from a different perspective, but this time we realize that not only are we not hearing the events "objectively" but there had been a gap in information before. As theorist Mieke Bal has observed, events presented in more than one light through internal retroversion like this are "both identical and different" in the course of the narrative.[65] Enough retroversion, though, and the very notion of the availability of a stable linear time outside of the self, of moving toward security and longevity, erodes. The temporality Kaitlin is working toward with Charlie is the very thing the sequencing of the plot is working against. In this way, the piece follows a failed attempt to put a heterosexual mode of temporality and behavior into a queer time.

Narrative structure is, however, only one way of engaging with the work. Another productive way to listen to *The Shadows* is to ignore the words entirely and focus on breathing, on sounds that come from the ecstatic gut, the mucous-covered glottis, the groggy larynx, the sloppy tongue, and especially the expressive nose. This brings us back to Barthes's ideas on the voice mentioned in chapter 1, only in this case with a focus on the erotic dimension of the "grain" of the voice. It was, after all, an effort to understand his own excitation by one voice over another that Barthes developed his "bodily" approach to music, neglecting the adjectives we typically use to admire song and instead seeking out a noun, "the body in the voice as it sings," as the mark of materiality that stands outside of language and leads to jouissance.[66] Ask not whether the singer's voice is on pitch, if it is voluptuous or clear; instead, this approach goes, ask how many teeth she has, how much she drinks, what the shape of her vocal folds is, what the length of the frenulum connecting her gums to her lips is, because doing so leads to a sense not of virtuosity but of presence. Barthes's approach is compelling and paradoxical because it is an unashamed eroticism of the voice, subjectivity itself, while at the same time it insists on "facts" about the voice, objectivity itself.

Prest's work really takes this insight to heart, focusing on the resonances of faces and bodies. Charlie's first appearance is through the sound of exhalation, as he rehearses a puppet show with Kaitlin at her mentor's studio. They pant together before they speak together. This is more than just noise or a space between dialogue. It's a story told by bodies. Charlie is a nose breather. Kaitlin favors her mouth and has high whispery tones that fill the whole soundscape of the series. Devon's voice is silky smooth, full of lung; when he and Kaitlin meet on a train in Newfoundland, they do little more than sigh, one after the other. When Kaitlin dumps him, the sighing Devon is reduced to a snotty, sniffly mess. The expressivity of these recorded noses and mouths gradually diminishes over the piece: by the end of Kaitlin and Charlie's relationship, we hear less and less of the facts of their faces as they face up to the facts.

Note that we hear these bodies from extremely close. The technique is distinct from most audio drama, where we are often encouraged to imaginatively position upright bodies in larger space, like figures on a chessboard, which is why the emphasis is usually on footsteps, doorways, and other thresholds. *The Shadows*, by contrast, doesn't need to communicate distant coordinates or fast choreography of heroes chasing villains down endless alleyways. Everyone the story needs is nearby. Indeed, although there are location markers—the restaurant where Kaitlin works, the sound of a drum circle at a famous Montréal monument, the signature screech of the Toronto subway—for the most part the kinetic world of the play happens inches from the bodies of its characters or closer. *The Shadows* tells us about its characters by how they breathe near one another microsonically, not by how they walk around one another. That's why the series is at its most "visual" not on the puppetry stage but in bed. In creating the piece, Prest and Wang opted mostly for location recording, which was done with inexpensive Zoom H1 stereo recorders held by the actors themselves. The technique has obvious ramifications for the "grain" of the whole series, as it picks up all manner of sonic detritus, plosives, muttering, and sniffling, words and phrases whispered so low you nearly miss them. Even mic-handling noise is not discarded, lending the piece's most intimate moments of pillow talk the feeling of cinéma vérité. The effect is like turning distal stimulus into proximal stimulus. *The Shadows* may be the only audio play that gives the illusion of having its own odor.

The erotic sloppiness of the sound capture technique is balanced in *The Shadows* by the sharpness of its editorial technique, with its use of extended runs of fast scenes that come across as flurries of working, eating, walking, playing, and having sex that rapidly change scene, format, and tense. When Kaitlin and Devon have their tryst, we hear a sex scene, followed by a singing scene, followed by another sex scene, followed by bedroom banter, then more sex, then a call to Charlie, all in about a minute. These rapid sequences expand the world of the piece. Note the use of the voices of Prest's real and imagined friends in her head—an internalized kaleidosonics—throughout *The Shadows*. Her mind is never quiet, never alone; it is a whole village that unfurls in musically phrased sequences that move to a thudding, intensifying beat. The technique lends the show flow, texture, and movement. It densifies the work and speeds up time.

Contrast that strategy with another that we also hear in *The Shadows*, one based on shock and reversal. The first episode begins with an instance of this. Kaitlin asks about how love ends and begins, a sequence punctuated with sobbing scenes of breakup, but then quickly segues to a young baby crying, then a chugging hardcore song, with hoarse lyrics ("A new life begins . . ."). The sequence disturbs the play's tone. Is this supposed to be touching, frank, funny? It's unclear. Other reversals are more jarring. Episode 2 starts with a long sequence of Charlie's fantasized headlines about future fame as a puppeteer, which feeds into an imagined interview with NPR icon Terry Gross. But it's the real Terry Gross, playing herself, that sends a Brechtian ripple across the texture of the story. Another alienation occurs when a scene of stifled arguments between Kaitlin and Charlie ("we're fighting over what we've given up over one another") segues to the real Kaitlin's interviews with her real parents about the happiness of their marriage. In episode 6, Kaitlin's reverie about what it felt like to masturbate as a young Catholic girl ("ashamed, disgusted, and confused") is cut off by uncontextualized vomiting; it takes several moments until we learn that we are in the present and Kaitlin is at a wedding in the Catskills, hurting after too much booze.

Smooth kaleidosonic sequences and shock/reversal editing choices are just a backdrop to an even more defining method of multiplicity. There are several points in the narrative in which two temporalities of voice—character narrator and character in scene—overlap with a third. When Kaitlin and Charlie kiss for the first time, for example, we hear three separate Kaitlin voices speaking to us, a narrator, the character, and a "second-self." Later, sixteen minutes or so into episode 4, as Kaitlin debates a possible future with Devon and one with Charlie, I count twenty-nine separate overlapping phrase fragments emerging in just sixty seconds, distributed in various places across the width of the stereo image as Prest accuses, absolves, denigrates, and vindicates herself in a maddeningly recursive loop. Here is just a fragment:

"Everything would be fine if I could just ask Charlie to be in an open relationship."
"Yeah, but would it really?"
"My obsession with Devon feels like a full-time job . . ."
"While I'm working the other full-time job which is spending as much time as humanly possible naked with Charlie . . ."
"While I'm working my part time job of waitressing . . ."
"While I'm working—"
"The job that I've always wanted"
"—which is having actual real puppetry contracts"
"Like love might not be finite, but the hours of the day are . .

The sequence pops out like spattered paint on a canvas as it devolves into self-frustration. And the fragmentation it emphasizes persists even after she makes a decision:

KAITLIN AS NARRATOR: "In the end, I decide, I'm never going to tell Charlie."
KAITLIN ON THE RIGHT STEREO CHANNEL: "It's *your* body, *your* heart."
KAITLIN ON THE LEFT STEREO CHANNEL: "That is so dishonest and fucked up."
CHARLIE: "What's wrong?"
KAITLIN: "Nothing."

In this work of audio fiction, editing multiplies difference within voices and, by correlation, within characters. Consciousness is chaos in stereo space. I am reminded of Gregory Whitehead's notion of how to approach the voice in radio. "The fact is, we cannot find our voice just by using it," he writes. "We must be willing to cut it out of our throats, put it on the autopsy table, isolate and savor the various quirks and pathologies, then stitch it back together and see what happens. The voice, then, not as something which is found, but as something which is written."[67]

The voice is but one of several fetish objects *The Shadows* uses as an orienter. In her classic book *Queer Phenomenology*, Sara Ahmed writes of how our understandings of sexuality are bound up in our understanding of positionality, of "orientation" and all that word implies. This idea leads her to an engagement with the phenomenology of Husserl, considering how seemingly inert objects play a role in perception. "We are turned toward things. . . . Things make an impression upon us," Ahmed writes. "We perceive them as things insofar as they are near to us insofar as we share a residency with them. Perception hence involves orientation; what is perceived depends on where we are located, which gives us a certain take on things." For Ahmed, objects give us orientation in the world, just as our particular orientation in the world gives objects their status as objects relative to us.[68] Bear that in mind when you consider Sweater, a speaking object passed from Devon to Kaitlin when they first meet. Sweater becomes the star of the surrealist third episode and perhaps the only really likeable character in *The Shadows*.

Of course, Sweater is not the only fetishized object in the show. Charlie gets obsessed with cocooning his finger with Kaitlin's finger through wax; Kaitlin's love letters to Devon insist that "I want you in my toothpaste." The microphone itself is another. In a scene at the airport, we hear the mic crushed between Devon's and Kaitlin's bodies, a scene that oddly gives us great intimacy and great distance at once, as the mic struggles to connect their two heartbeats. All of these objects reflect orientations of desire; it is as though desire couldn't exist without them, the occlusions they offer, the intermingling they manage. On the subject of skin and touching objects, Ahmed writes, "We perceive the object as an object, as something that 'has' integrity and is 'in' space, only by haunting that very space; that is, by co-inhabiting space such that the boundary between the co-inhabitants of that space does not hold. The skin connects as well as contains."[69]

Still, Sweater is the most significant orienting presence in *The Shadows*. In monologues, it knows systems of exchange and gift ("the lending of a sweater is a sacred message"), it knows where it comes from (a sheep named Gladys), it knows what love is ("providing a barrier between the tender skin and the winds of the cruel city"), and it knows what it wants, the coveted place of "favorite" sweater, the love of its owner. The fiction does all it can to mark Sweater with abject fluids—sweat, tears, saliva, snot, ketchup, sexual fluids—to make it at once disgusting and wealthy with material density. And *The Shadows* also does all it can to give Sweater as rich of an imaginative life as anyone else in the story, with lovely passages of sweater sex and fantasies of marrying Kaitlin, in the first human-sweater relationship sanctioned by law. By the end of the episode, all this changes. Kaitlin has scorned Devon and elected to wash all the effluvial humors associated with him from Sweater, which is now again diminished from its role as an object of love and revealed instead to be a mere mediator of love between others. We hear a series of tortures: the long-unwashed sweater having its memories dissolving in suds; an amusing but destructive tumble-dry sequence that oscillates between our earbuds; and, overtop of melodramatic strings, in the single most closely miked detail in the entire series, the sound of moths chewing it up. It's like a jealous ritual designed to take Sweater from us as listeners, a literal disorientation of the listener from the world of the fiction and the object through which we achieved residence within it.

All the features I've been describing—the mood of cruel optimism, prominence of queer time, micro-sonics and bodily sound, the emphasis on editing technique and on an orientation toward objects—make *The Shadows* perhaps the one widely known work of fiction in this chapter that openly sought to make a great leap forward in audio aesthetics when it comes to drama. In this case, anamnesis through classical radio drama seems to be less of a critical priority, and yet it is striking how often the piece takes on an atavistic feel, constantly pointing to the past. *The Shadows* starts and ends after all with the creation of a classical performance. In the beginning, Kaitlin dresses as a mermaid and sings a plaintive song with her accordion in the subway. It is the siren song that brings Charlie to her, that turns her into a love object. "Do you know how you fuck a mermaid?" she asks, as their relationship deepens. "You don't fuck a mermaid. A mermaid fucks you." Six episodes later, that's just what's happened, and the series ends at the same setting. She has finally ended the relationship, got good and drunk while staring out at sailboats, and then taken up the mermaid outfit again, heading underground to perform as the "heartbreak mermaid," a human puppet, a siren returning to her rock. She sings wordlessly, the unlubricated keys of her old accordion thunking along as accompaniment, for a minute and a half. Then the sound bleeds into the screech of subway cars, and a dusty record plays us out. In Prest's work, the incompatibility of the passion for feeling love and the zeal for deconstructing it often produces scenes of surprising angst uncommon in modern audio.

What to make of this tableau, of Kaitlin saying goodbye not to the object of her love but to the process of saying goodbye to love, particularly as it takes place inside a brazen allusion to the *locus classicus* of sound art, the forbidden song of the Sirens in *The Odyssey*, an embodiment of nostalgia and the hidden knowledge of the forgotten past? In truth, *The Shadows* has been looking for just such a moment of exquisite, suspended animation all along. When Kaitlin seeks out advice from a woman who had an affair for eighteen years, she discovers the older woman to be paralyzed by the choice of whether to visit her longtime side lover, who is dying, despite her husband's wishes. When Kaitlin tries to understand how her parents love one another, she finds herself dumbfounded by a home movie of the pair reacting to one another with unexpected tenderness. Similar tableaux occur in her own love story, particularly in fantasies that dominate so much of the imaginative life of the series—fantasies of artistic glory, of aging, of gardening with a lover, of travel, of being naked in front of her children, of righteous celibacy. Kaitlin tries on a procession of idealized puppet selves, possible futures that shimmer all the brighter because they are unchosen. I am reminded of Svetlana Boym's insight that nostalgia isn't always about the past; it can be prospective, too. In this moment, Kaitlin feels mournful not for her connection to lived selves but for her connection to possible lived selves she has abandoned.

Indeed, it is striking that in this most sonically original of audio dramas, the one that experientially seems to step into its own retroverted, visceral, garrulous aesthetic ends in an unsatisfied, recumbent quiet. Kaitlin in her subway is Boym's nostalgist par excellence, longing for something that never existed, trapped by time and space. Although the stories we find them in couldn't be more different stylistically, Kaitlin in *The Shadows* and Heidi in *Homecoming* are in a similar predicament, unable to come home to something but unable not to come to terms with it, surviving unfree of their pasts. Maybe this captures something about the situation of audio drama in its most recent form, too. Were dramatists ever to step entirely outside of the perimeter of the aesthetics of amnesia to invent audio drama entirely anew, perhaps it would all still sound like a memory, anyway. On this level, *The Shadows* reminds us that nostalgic thinking was once considered to be not a cultural nourishment or indulgence but a sickness. One reason amnesia is so attractive as an aesthetic mode, after all, is that it represents the protection from memory, even imagined memory, lest it become a paralyzing affliction.

In her reading of *The Odyssey*'s Siren episode, Boym observes that "modern nostalgia is a mourning for the impossibility of mythical return, for the loss of an enchanted world with clear borders and values."[70] In the face of that loss, nostalgists double back, get lost, move in zigzags. They never make it home. Instead, Boym writes, "The nostalgic is looking for a spiritual addressee." So is audio fiction, a genre forever searching for those willing to subscribe to its cherished myth that there are poignant singular moments of sonic experience that stand outside of memory, like the moment a siren begins to sing, not for sailors on the waves but for herself.

A Return to the Memory Hole

In the spring of 2022, less than seven years after the Third Coast Filmless Festival session that I described at the start of this chapter, I found myself once again in a large room in a rented high school. This time it was an auditorium in the Bronx. I was watching a group of artists working on a radio play, but I was on a film set rather than at a filmless festival. I'd been hired to consult for a movie sequence that included a 1930s radio play, for which I was asked to give advice on art direction for props and sound effect schemes, to help make the scene look and sound authentic. After spending much time thinking about audio drama for the twenty-first century while writing this book, it was refreshing to be looking at a radio drama for the twentieth century, one set in the past entirely without irony, taking seriously the problem of how to re-create a moment when audio drama really was "new" and professionalizing for the first time, in a scene that was not reduced to the winks and smug sentimentality that modern re-creations of old-time radio so often foreground. The stage was complete with almost everything that "modern" audio dramas wanted to hide about their long ancestry: a costumed 1930s band, transcription disc players for sound effects, old chains for a prison sequence, and a starter pistol onstage, along with an announcer and actors reading scripts with curled corners for easier page turning in front of vintage microphones. It was the very world of narrative audio that modern podcasters had spent so much energy disavowing in recent years, as it is almost never seen: unsmirkingly, without apology.

The shoot took place shortly after the Omicron COVID wave, toward the end of the main phase of the pandemic in the United States, at least as of the time of writing. Many figures at the center of audio drama in the 2010s were at that time still working in the field, though the various lanes separating prestige audio, indie work, audiobook projects, and podcast company series as I set them out above—never tidy to begin with—had become entirely blurred. Ann Heppermann became an executive producer at Audible in 2020, supervising a variety of projects; Third Coast went entirely online and would not return to in-person events until the summer of 2023, as new venues such as the Tribeca Festival emerged to showcase narrative audio; the Sarah Awards, along with *Serendipity*, seemed to peter out. Independent producers continued to have viral hits, like Tal Minear's *Someone Dies in This Elevator*. Veterans became interested in limited anthologies, like Sarah Shachat, Zach Valenti and Gabriel Urbina's *Zero Hours*, a mediative and often theological exploration of the concept of the end of the world as told in conversational scenes of characters in dyads—a minister and a witch in eighteenth-century New England; a scientist and an overlord centuries in the future—trying to snooker one another in the face of apocalypse. Mac Rogers and his collaborators released *Give Me Away*, a fascinating series about a man volunteering to let his brain host a disembodied alien mind as part of his own self-therapy. *The Truth* continued its series of experiments in metathrillers and comedies, before eventually closing its doors in 2023 with "Pariah," an outstanding series on art and its limits.

Mid-sized companies and networks continued to emerge, such as Cadence 13, producer of the full-length piece *Ghostwriter* with Kate Mara, which reminisced the audible airport thrillers of eight years before, and Fable and Folly, which represented prominent independent audio dramas to potential advertisers. The Dirk Maggs / Neil Gaiman collaboration produced sophisticated, long-form adaptations of Gaiman's famous comics piece, *The Sandman*, perhaps the one thing for which this rewarding collaboration will be remembered in future years; for complexity these shows work at another level, and its narrative takes more risks than expected—one chapter features both a necrophilia fantasy and a musical number. Kaitlin Prest continued her experiments in both fiction and nonfiction, founding an art company called Mermaid Palace, which produced several new experimental pieces of semi-fiction exploring themes of family, queerness, and violence, including Sharon Mashihi's *Appearances* and Drew Denny's *Asking For It*. In an article on the new company at its launch, Elena Fernández Collins described Prest's recent role as a "doula for other people's voices."[71] Besides Mermaid Palace, the BBC also took a role pushing experimental literary approaches through *Have You Heard George's Podcast?*, an amalgam of spoken word and the audio feature, and *Forest 404*, a thriller about an archive of soundscape recordings that contains lectures and artist renderings about its sounds.

Gimlet Media continued its expansion into fiction following its purchase by Spotify, with multi-season shows such as *Motherhacker*, by Sandi Farkas and Amanda Lipitz, starring TV stars Carrie Coon, Pedro Pascal, and others. Soon the company would release *Case 63*, which saw Oscar Isaac return to the role of a psychiatric patient once more, this time in a show adapted from *Caso 63*, a top podcast in Argentina and Mexico that was adapted to English by Mara Vélez Meléndez and directed by Mimi O'Donnell about a time traveler from the future coming to (apparently) stop a pandemic, a story reminiscent of Chris Marker's famous 1962 film

La Jetée. Eli Horowitz, director of *Homecoming*, followed in the footsteps of Paul Bae, Mischa Stanton, Lauren Shippen, and Brendan Baker into the world of superhero podcasts, making 2023's *Harley Quinn and The Joker: Sound Mind* for the DC universe, collaborating with Fred Greenhalgh of *The Cleansed*. It was the second Spotify-DC collaboration in the Batman universe, both of which briefly unseated Joe Rogan at the top of the podcast charts in the United States (and had similar success around the world), and drew in their scenography and premises liberally from 1990s horror films like *The Silence of the Lambs* and *Se7en*.

There were also new developments, some of which were reactions to the lockdown itself. With many theaters going dark, various stage drama groups began to experiment with audio formats. New York's Playwrights Horizons commissioned a series called *Soundstage*, with a number of experimental historical and contemporary pieces released online, eventually creating a full-blown series about the revival of the cruise industry, while the Public Theater distributed an audio version of Shakespeare's *Richard III*. The La Jolla Playhouse created a four-part radio horror show, *Listen with the Lights Off*, while the Rattlestick Playwrights Theater made a series called the *MTA Radio Plays*, an anthology of short pieces by Ren Dara Santiago set along a trainline from the Bronx to Brooklyn. In the United Kingdom a number of dramatists brought old and new plays to the podcast space, such as Simon Stephens performing José Saramango's novel *Blindness* and Mike Bartlett's satire *Phoenix*, written, rehearsed, and performed in ten days. Meanwhile, by 2022, many of Audible's investments in theater projects had begun to come to fruition, and the brand could boast some ninety-three "theater" productions, from newer plays like Adam Rapp's *The Sound Inside* and Madhuri Shekar's *Evil Eye* to adaptations of classic works by Tennessee Williams, Joan Didion, and Eugene O'Neill, some of which the brand put on stage before committing to audio and others that were drawn from prominent festivals.[72] During the pandemic, Audible took on producing a whole season of the Williamstown Theatre Festival, rehearsing and recording the shows the group was unable to present.

Another major player on the scene in these years was QCode. After success with the tape-based post-disaster thriller *Blackout*, the distributor began releasing a deluge of titles, many of them featuring celebrities and bearing some obvious connections to pre-pandemic audio. Its works had some familiar themes and devices. *Dirty Diana*, a series about recorded erotic fantasies starring Demi Moore as a finance executive in an unhappy marriage, used the same "sonic mezzanine" that *Limetown* had used for its stories-within-stories. Nia DaCosta's *Ghost Tape* brought elements from both the psychiatric and tape-based fetishes that audio dramas had been exhibiting since the mid-2010s. In a review of QCode's *Left-Right Game*, critic Rashika Rao pointed out that some of the company's shows seemed to resemble Joseph Fink's *Alice Isn't Dead* by accident, indicating a troubling "lack of familiarity with the existing landscape" by creators.[73] Founded by a former Creative Artists Agency talent agent, Rob Herting, QCode set out to generate intellectual properties he could sell to other mediums, hoping to churn out twenty or so a year, trying to scale up a model that was pioneered by indie YouTube and web TV series.[74] Making audio drama was approached essentially as a way to make movies and TV on spec.

Noting these developments that fall, newcomer critics to audio drama began asking questions about what they ought to call these "new" works. In an October 2021 *Variety* interview about this "new frontier" for podcasts, Rob Herting is described as "tossing around" terms like "audio drama" and "movies for your ears."[75] Two months later a *New York Times* article proposed terms like "podcast movie" and "feature-length audio movie," for what it imagined to be a brand-new medium, citing works like *Ghostwriter*.[76] The *Times* article received negative attention from the established audio drama podcast community, with *Forbes* pointing out that many felt that this sudden new interest "demean[ed] or ignore[d] the work they ha[d] done for years in the audio fiction space" and quoting *The Truth*'s Jonathan Mitchell, who described the new discourse of next-big-thingness as "straight up offensive."[77] It was as if the entire body of podcasts described in this chapter had already begun to vanish into a memory hole.

If this chapter has proven anything, it is that sudden memory loss like this is inevitable, as it has long been the genre's habit to undergo it. But that doesn't diminish the impact of how suddenly a past can vanish. Perhaps it is the destiny of all books like this one that begin by overcommitting to a particular temporal framing—my insistence on a retrospective mood rather than a proleptic imaginary—that they cannot help but unravel in the end. Watching the 1930s radio play being re-created in front of me, another birth story of the medium, and reading about these developments that spring, I had the feeling of being caught in the amnesiac's perpetual present. The whole cycle of amnesia and anamnesis was beginning again, only this time the dramatists who pushed life into the medium in the 2010s were seeing the turn of the wheel from its other side. Indifferent to efforts at making the case for clarifying retrospection that might slow the cycle of perpetual rebirth and finally bring the medium to permanence and offer an edifice of contiguity, audio drama had returned once more to the unforgiving place it is always found, a medium at once long gone yet also forever at the moment of its birth, about to come into replete existence, any day now.

Audio Works Discussed in Order of Appearance in the Text—Born Podcast/Hybrid

Serendipity (Ann Heppermann, Martin Jonson, The Sarah Awards, July 3, 2015–June 15, 2017, https://www.wbez.org/shows/serendipity/a3958ae0-a6bc-4af9-901f-c6598f853d16).

Serial (Sarah Koenig, WBEZ Chicago, October 3, 2014–present, https://serialpodcast.org/).

L.A. Theatre Works (Susan Albert Lowenberg, LATW, February 9, 2017–present, https://latw.org/podcasts).

Welcome to Night Vale (Jeffrey Cranor, Joseph Fink, Night Vale Presents, June 15, 2012–present, https://www.welcometonightvale.com/).

The Truth (Jonathan Mitchell, Radiotopia, February 11, 2012–present, http://www.thetruthpodcast.com/).

The Heart (Kaitlin Prest, Mermaid Palace, Radiotopia, February 5, 2013–February 23, 2022, https://www.theheartradio.org/all-episodes).

Radiolab (Jad Abumrad, Robert Krulwich, et al., WNYC, May 25, 2012–present, https://radiolab.org/episodes).

This American Life (Ira Glass, WBEZ Chicago, November 17, 1995–present, https://www.thisamericanlife.org/archive).

Radio Diaries (Joe Richman, Radiotopia, May 16, 2013–present, https://podcasts.apple.com/us/podcast/radio-diaries/id207505466).

Invisibilia (Lulu Miller, Alix Spiegel, NPR, January 9, 2015–present, https://www.npr.org/programs/invisibilia/).

Criminal (Phoebe Judge, Radiotopia, January 28, 2014–present, https://thisiscriminal.com/).

Mystery Show (Starlee Kine, Gimlet Media, May 21, 2015–July 31, 2015, https://gimletmedia.com/shows/mystery-show/episodes#show-tab-picker).

The Moth Radio Hour (Jay Allison, Meg Bowles, George Dawes Green, Jenifer Hixson, Suzanne Rust, The Moth, Atlantic Public Media, Public Radio Exchange, 2009–present, https://themoth.org/radio-hour).

Snap Judgment (Glynn Washington, Snap Judgment Studios, PRX, July 2010–present, https://snapjudgment.org/podcast-episodes/).

Limetown (Zack Akers, Skip Bronkie, Two-Up Productions, July 29, 2015–December 3, 2018, https://twoupproductions.com/limetown/podcast).

The Far Meridian (Mischa Stanton, Eli Barraza, The Whisperforge, June 5, 2017–May 25, 2020, https://www.whisperforge.org/thefarmeridian/season-1).

Unwell: A Midwestern Gothic Mystery (Jeffrey Gardener, Eleanor Hyde, Jim McDoniel, HartLife NFP, February 20, 2019–present, https://www.unwellpodcast.com/s1e1-homecoming-1).

The Bright Sessions (Lauren Shippen, Atypical Artists, November 1, 2015–November 24, 2021, https://www.thebrightsessions.com/listen).

Homecoming (Micah Bloomberg, Eli Horowitz, Gimlet Media, November 16, 2016–November 2, 2018, https://gimletmedia.com/shows/homecoming/episodes).

Ghost Tape (Nia DaCosta, QCode, October 26, 2020–December 14, 2020, https://podcasts.apple.com/us/podcast/ghost-tape/id1536055234?ls=1&itscg=30200&itsct=qcode_podcasts&at=1001l36mQ&ct=GhostTape-Web).

Batman: Unburied (David S. Goyer, Phantom Four & Wolf at the Door in association with Blue Ribbon Content and DC for Spotify, April 2022–May 2022, https://open.spotify.com/show/3pUWoZ6fC2qA02D3X0CeMb).

The Message (John Dryden, GE Podcast Theater, Panoply, October 3, 2015–November 21, 2015, https://podcasts.apple.com/us/podcast/lifeafter-the-message/id1045990056).

Bellwether (Sam Greenspan, September 20, 2021–October 11, 2021, https://podcasts.apple.com/us/podcast/bellwether/id1471742168).

Blackout (Scott Conroy, Rami Malek, Endeavor Content, QCode, March 19, 2019–October 7, 2021, https://podcasts.apple.com/us/podcast/blackout/id1447513097?ls=1&itscg=30200&itsct=qcode_podcasts&at=1001l36mQ&ct=Blackout-Web).

Case 63 (Mimi O'Donnell, Mara Vélez Meléndez, Katie Pastore, Gimlet Media, October 25, 2022, https://gimletmedia.com/shows/case-63).

Mabel (Becca De La Rosa, Maybell Marten, September 30, 2016–April 15, 2022, https://mabelpodcast.com/episodes).

Within the Wires (Jeffrey Cranor, Janina Matthewson, Night Vale Production, June 20, 2016–present, http://www.nightvalepresents.com/withinthewires).

Archive 81 (Daniel Powell, Marc Sollinger, Dead Signals, April 5, 2016–June 4, 2019, https://podcasts.apple.com/us/podcast/archive-81/id1098194172?mt=2).

The Magnus Archives (Lowri Ann Davies, Alexander J Newall, Jonathan Sims, Rusty Quill, March 23, 2016–November 4, 2021, https://rustyquill.com/show/the-magnus-archives/).

Video Palace (Nick Braccia, Mike Monello, Ben Rock, Shudder, October 16, 2018–October 26, 2020, https://open.spotify.com/show/4u74IOaFUCejkHsH8sozCd).

Darkest Night (Lee Pace, The Paragon Collective, Shudder, October 31, 2016–November 28, 2018, https://soundcloud.com/darkestnightpod).

We're Alive (Kc Wayland, Wayland Productions, Inc., May 4, 2009–November 5, 2019, https://www.werealive.com/listen/).

What's the Frequency? (Alexander Danner, James Oliva, September 13, 2017–September 11, 2018, https://wtfrequency.com/category/audio/).

The Shadows (Kaitlin Prest, Phoebe Wang, CBC Podcasts, September 24, 2018, https://podcasts.apple.com/us/podcast/the-shadows/id1420121326).

The Big One (Misha Euceph, Arwen Nicks, KPCC, LAist Studios, January 10, 2019–July 22, 2019, https://laist.com/podcasts/the-big-one).

Price of Secrecy (Zoha Zokaei, Radio Atlas, August 5, 2019, http://price-of-secrecy.com/).

The Leviathan Chronicles (Christof Laputka, Leviathan Audio Productions, April 20, 2008–August 15, 2022, https://www.leviathanchronicles.com/episodes/).

The Cleansed (Frederick Greenhalgh, FinalRune Productions, WMPG, March 9, 2012–September 25, 2016, http://thecleansed.com/downloads/).

The Behemoth (Rick Coste, March 1, 2017–March 20, 2017, https://modernaudiodrama.com/the-behemoth.html).

A Night Called Tomorrow (James Urbaniak, HowlFM, 2017, https://www.stitcherpremium.com/night).

Campfire Radio Theater (John Ballentine, A Haunted Air Audio Drama, December 17, 2011–October 23, 2022, https://campfireradiotheater.podbean.com/).

Our Fair City (Clayton Faits, Jeffrey Gardner, HartLife NFP, March 28, 2012–July 4, 2018, https://podcasts.apple.com/us/podcast/our-fair-city/id514748675).

Transmissions from Colony One (John Richter, John Richter Creative, June 20, 2022–present, https://www.tfco.us/listen).

Wolf 359 (Zach Valenti, Gabriel Urbina, Sarah Shachat, Kinda Evil Genius Productions, August 15, 2014–December 24, 2017, https://wolf359.fm/season1).

Tin Can (David Devereux, Tin Can Audio, July 3, 2016–May 6, 2018, https://www.tincanaudio.co.uk/tincan).

The Bridge (Alex Brown, Rebecca Mahoney, The Bridge Podcast, July 20, 2016–February 23, 2021, http://www.thebridgepod.com/episodes).

The Orphans (James Barbarossa, Zachary Fortais-Gomm, Ella Watts, The Light and Tragic Company, May 23, 2016–present, https://link.chtbl.com/d9niQcvg?sid=LTWEB).

The Black Tapes (Paul Bae, Terry Miles, Pacific Northwest Stories, May 21, 2015–January 12, 2020, http://theblacktapespodcast.com/).

Rabbits (Terry Miles, Public Radio Alliance, February 28, 2017–March 15, 2022, https://www.rabbitspodcast.com/episodes).

TANIS (Terry Miles, Public Radio Alliance, October 13, 2015–November 18, 2020, http://tanispodcast.com/).

The Big Loop (Paul Bae, QRX, October 31, 2017–October 23, 2018, https://www.thebiglooppodcast.com/).

Tumanbay (John Scott Dryden, Mike Walker, Goldhawk Productions for BBC Radio 4, Panoply, https://www.tumanbay.com/season-one).

Marvel's Wolverine: The Long Night (Brendan Baker, Daniel Fink, Jenny Radelet, Marvel, Sirius XM, September 12, 2018–November 7, 2018, https://www.wolverinepodcast.com/episodes-season-1/).

Lif-e.af/ter (Mac Rogers, GE Podcast Theater, Panoply, November 12, 2016–January 28, 2017, https://podcasts.apple.com/us/podcast/lifeafter-the-message/id1045990056).

Steal the Stars (Mac Rogers, Gideon Media, August 1, 2017–October 31, 2017, https://soundcloud.com/user-965073203).

Louis L'Amour (various) (Penguin Random House Audio, https://www.penguinrandomhouseaudio.com/author/16445/louis-lamour/).

Alice Isn't Dead (Joseph Fink, Night Vale Production, March 7, 2016-November 14, 2019, http://www.nightvalepresents.com/aliceisntdead).

Audio Drama Production Podcast (UberDuo Podcast Network, September 16, 2014–November 17, 2019, https://open.spotify.com/show/3gqGRI1OGU6pDxjLAzeFsB).

Radio Drama Revival (Fred Greenhalgh, January 22, 2007–present, https://radiodramarevival.com/).

The X-Files (Fred Greenhalgh, Audible Originals, March 12, 2019, https://www.amazon.com/X-Files-Cold-Cases-Joe-Harris/dp/1978665253).

Bronzeville (Lawrence Fishburne, Larenz Tate, Kc Wayland, Cinema Gypsy Productions, TateMen Entertainment, Audio HQ, DAX US/Global, February 7, 2017–April 20, 2021, https://podcasts.apple.com/us/podcast/bronzeville/id1199964972).

Valence (Anne Baird, Katie Youmans, Wil Williams, Hug House Productions, January 11, 2020–present, https://open.spotify.com/show/3OfZe1ByjKWAlynyqjJcmJ).

Someone Dies in this Elevator (Tal Minear, Realm, April 26, 2021–present, https://open.spotify.com/show/5mTHlUvloBQ6KfqnmJuOn7).

The Starling Project (Jeffrey Deaver, Audible Originals, May 26, 2015, https://www.amazon.com/The-Starling-Project-audiobook/dp/B00OZ4UGQK).

The Child (Sebastian Fitzek, Audible Originals, December 17, 2015, https://www.amazon.com/The-Child-Sebastian-Fitzek-audiobook/dp/B00LETG46U).

Locke & Key (Joe Hill, Audible Originals, October 5, 2015, https://www.amazon.com/Locke-Key-audiobook/dp/B010PNSBYI).

Alien: Out of the Shadows (Tim Lebbon, Dirk Maggs, Audible Studios, June 9, 2016, https://www.amazon.com/Alien-Out-of-Shadows-Tim-Lebbon-audiobook/dp/B01ENIY7PI).

Blood on Satan's Claw (Piers Haggard, Mark Morris, Robert Wynne-Simmons, Bafflegab Productions, Audible Studios, January 16, 2018, https://www.amazon.com/Blood-on-Satans-Claw-audiobook/dp/B078J9GH76).

Lincoln in the Bardo (George Saunders, Penguin Random House Audio, February 14, 2017, https://www.penguinrandomhouseaudio.com/book/231506/lincoln-in-the-bardo/).

Girls and Boys (Dennis Kelly, Lyndsey Turner, Audible Originals, June 27, 2018, https://www.amazon.com/Girls-Boys-Dennis-Kelly-audiobook/dp/B07DKP97LP).

The Thrilling Adventure Hour (Ben Acker, Ben Blacker, The Forever Dog Podcast Network, January 23, 2011–present, https://podcasts.apple.com/us/podcast/the-thrilling-adventure-hour/id408691897).

Comedy Bang! Bang! (Scott Aukerman, Earwolf, Midroll Media, May 1, 2009–present, https://www.earwolf.com/show/comedy-bang-bang/).

Crimetown (Marc Smerling, Zac Stuart-Pontier, Gimlet Media, 2016–19, https://gimletmedia.com/shows/crimetown/episodes#show-tab-picker).

Stranglers (Portland Helmich, Earwolf, Northern Light Productions, November 16, 2016–February 15, 2017, https://podcasts.apple.com/us/podcast/stranglers/id1174116487).

Unsolved Murders: True Crime Stories (Wenndy Mackenzie, Carter Roy, Parcast,

Spotify, June 4, 2016–present, https://open.spotify.com/show/6Mk7Wk6cmrgXWnitO3NdYe).

Hollywood & Crime (Stephen Lang, Tracy Pattin, Wondery, January 6, 2017–November 3, 2017, https://wondery.com/shows/hollywood-crime/).

Inside Psycho (Mark Ramsey, Wondery, March 23, 2017–July 10, 2017, https://wondery.com/shows/inside-psycho/).

Secrets, Crimes & Audiotape (Hernan Lopez, David Rheinstrom, Wondery, September 27, 2016–October 24, 2017, https://wondery.com/shows/secrets-crimes-audiotape/).

The Mysterious Secrets of Uncle Bertie's Botanarium (Chris Bannon, Fiona Elwood, Duncan Sarkies, Stitcher, South Coast Shenanigans, March 15, 2016–April 18, 2017, https://www.stitcher.com/show/the-mysterious-secrets-of-uncle-berties-botanarium).

Hunted (Dick Wolf, Wolf Entertainment, Endeavor Content, November 26, 2019–December 10, 2019, https://podcasts.apple.com/us/podcast/hunted/id1484428621).

Marvels (Lauren Shippen, Stitcher, March 18, 2020–May 20, 2020, https://www.stitcher.com/show/marvels).

Anthem: Homunculus (John Cameron Mitchell, October 13, 2021–November 10, 2021, https://www.topic.com/anthem).

The Passenger List (John Scott Dryden, Mark Henry Phillips, Lauren Shippen, Radiotopia, July 29, 2019–July 21, 2021, https://passengerlist.org/).

The Unexplainable Disappearance of Mars Patel (Benjamin Strouse, Pinna and Gen-Z Media, June 20, 2021–July 23, 2021, https://gzmshows.com/shows/listing/mars-patel/).

Wooden Overcoats (David K. Barnes, Andy Goddard, John Wakefield, September 23, 2015–March 31, 2022, https://www.woodenovercoats.com/listen).

Issa Rae Presents . . . Fruit (Issa Rae, Issa Rae Productions, Midroll Media, February 2, 2016–July 17, 2017, https://podcasts.apple.com/us/podcast/issa-rae-presents-fruit/id1256093382).

Dark Adventure Radio Theater (H.P. Lovecraft Historical Society, April 23, 2018–February 10, 2022, https://store.hplhs.org/collections/dark-adventure-radio-theatre).

A Court of Thorns and Roses (Duane Beeman, Anji Cornette, Sarah J. Maas, Matt Webb, Graphic Audio, March 8, 2022–February 1, 2023, https://www.graphicaudio.net/our-productions/series/a-court-of-thorns-and-roses.html).

The First Mountain Man (Duane Beeman, Anji Cornette, William W. Johnstone, Matt Webb, Graphic Audio, November 1, 2007–present, https://www.graphicaudio.net/our-productions/series/first-mountain-man.html).

The Stormlight Archive (Anji Cornette, Brandon Sanderson, Graphic Audio, March 21, 2016–December 3, 2021, https://www.graphicaudio.net/our-productions/series/stormlight-archive.html).

The White Vault (K.A. Statz, Travis Vengroff, Fool and Scholar Productions, October 2, 2017–October 31, 2022, https://thewhitevault.com/).

Greater Boston (Alexander Danner, Jeff Van Dreason, March 15, 2016–December 21, 2022, https://greaterbostonshow.com/?page_id=2634).

Ars Paradoxica (Daniel Manning, Mischa Stanton, June 1, 2015–June 6, 2018, https://arsparadoxica.com/season-1).

EOS 10 (Justin McLachlan, PlanetM, October 15, 2014–August 12, 2019, https://podcasts.apple.com/us/podcast/eos-10/id928069318).

Rose Drive (Raul Vega, Phantom Ape Productions, August 28, 2017–September 6, 2019, https://podcasts.apple.com/us/podcast/rose-drive/id1270885001).

Janus Descending (Jordan Cobb, No Such Thing Productions, October 27, 2018–December 30, 2022, https://www.nosuchthingradio.com/janus-descending-listen).

Harlem Queen (Yhane Washington Smith, Gabrielle Adkins, February 24, 2019–August 2, 2021, https://www.harlemqueen.org/podcast).

The Amelia Project (Philip Thorne, Øystein Ulsberg Brager, Imploding Fictions, November 16, 2017–November 18, 2022, https://ameliapodcast.com/listen-now).

The Off Season (Jonathan Mitchell, The Truth, October 1, 2018–October 22, 2018, http://www.thetruthpodcast.com/story/2018/9/30/the-off-season).

The Body Genius (Jonathan Mitchell, The Truth, April 3, 2019–May 29, 2019, http://www.thetruthpodcast.com/story/2019/4/3/the-body-genius).

The Horror of Dolores Roach (Aaron Mark, Gimlet Media, October 16, 2018–October 16, 2019, https://gimletmedia.com/shows/dolores-roach/episodes#show-tab-picker).

Sandra (Matthew Derby, Kevin Moffett, Mimi O'Donnell, Sebastian Silva, Gimlet Media, April 18, 2018, https://gimletmedia.com/shows/sandra/episodes#show-tab-picker).

Batman Unburied (David S. Goyer, Phantom Four, Wolf at the Door, Blue Ribbon Content, DC for Spotify, May 3, 2022–present, https://open.spotify.com/show/3pUWoZ6fC2qA02D3X0CeMb).

Harley Quinn and The Joker: Sound Mind (Frederick Greenhalgh, Eli Horowitz, Realm, Blue Ribbon Content, DC for Spotify, January 31, 2023, https://open.spotify.com/show/0LwynFnRBJBHaZRdkqC9fc).

Give Me Away (Cara Ehlenfeldt, Mac Rogers, Jordana Williams, Gideon Media, July 16, 2021–August 28, 2022, https://www.gideon-media.com/give-me-away).

Someone Dies in this Elevator (Tal Minear, March 24, 2021–present, https://sditepod.com/).

Ghostwriter (Chris Corcoran, Kate Mara, Adam Pincus, Adam Scott, Alix Sobler, Best Case Studios, C13 Features, December 5, 2021, https://www.c13features.com/ghostwriter/).

Zero Hours (Sarah Shachat, Gabriel Urbina, Zach Valenty, Long Story Short Productions, October 2019–November 2019, https://www.zerohourspodcast.com/).

The Sandman (Dirk Maggs, Neil Gaiman, Audible Originals, July 15, 2020–September 28, 2022, https://www.amazon.com/The-Sandman/dp/B086WQ7J62).

Appearances (Sharon Mishihi, Mermaid Palace, Radiotopia, September 29, 2020–October 27, 2020, https://mermaidpalace.org/Appearances).
Have You Heard George's Podcast? (George the Poet, BBC Radio, August 31, 2019–September 15, 2021, https://www.bbc.co.uk/programmes/p0791 5kd/episodes/downloads).
Forest 404 (Timothy X Atack, Becky Ripley, BBC Radio 4, April 4, 2019–May 29, 2019, https://www.bbc.co.uk/programmes/p06tqsg3/episodes/guide).
Asking For It (Drew Denny, Kaitlin Prest, Mermaid Palace, CBC Podcasts, February 23, 2020, https://podcasts.apple.com/us/podcast/asking-for-it/id1498213373).
Motherhacker (Sandi Farkas, Amanda Lipitz, Gimlet Media, November 13, 2019–June 7, 2021, https://gimletmedia.com/shows/motherhacker/episodes#show-tab-picker).
Caso 63 (Julio Rojas, Spotify, November 11, 2020–October 17, 2022, https://open.spotify.com/show/20ch3IIqtWSSM4nfy11ZzP).
Soundstage (New York's Playwrights Horizons, October 15, 2018–present, https://pod.link/soundstageph).
Richard II (Saheem Ali, The Public Theater, WNYC, July 13, 2020–July 16, 2020, https://publictheater.org/media-center/series/richard-ii/richard-ii/).
Listen with the Lights Off (La Jolla Playhouse, So Say We All, October 15, 2020–April 5, 2021, https://lajollaplayhouse.org/wow-goes-digital/listen-with-the-lights-off/).
MTA Radio Plays (Ren Dara Santiago, Rattlestick Playwrights Theater, December 14, 2020–May 12, 2021, https://www.rattlestick.org/stream-now).
Blindness (Simon Stephens, José Saramango, Donmar Warehouse, August 13, 2020–August 22, 2020).
Phoenix (Mike Bartlett, ETT, Headlong, November 5, 2020, https://ett.org.uk/watch-and-listen/phoenix/).
The Sound Inside (David Cromer, Adam Rapp, Audible Originals, March 18, 2021, https://www.amazon.com/The-Sound-Inside/dp/B08Y98QP41).
Evil Eye (Madhuri Shekar, Audible Originals, May 2, 2019, https://www.amazon.com/Evil-Eye-Madhuri-Shekar-audiobook/dp/B07QMZHDJH).
Williamstown Theatre Festival, 2020 Season (Williamstown Theatre, Audible Originals, December 3, 2020–April 8, 2021, https://www.audible.com/ep/williamstown?source_code=MRQOR24812162004K1).
Dirty Diana (Shana Feste, Demi Moore, QCode Media, July 13, 2020–August 17, 2020, https://podcasts.apple.com/us/podcast/dirty-diana/id1522113627?ls=1&itscg=30200&itsct=qcode_podcasts&at=1001l36mQ&ct=DirtyDiana-Web).

Audio Works Discussed in Order of Appearance in the Text—Born Radio/Recording

Under Milk Wood (Dylan Thomas, BBC, January 25, 1954).
Destination Freedom (Richard Durham, WMAQ, June 27, 1948–November 19, 1951).

The National Radio Theater of Chicago (Yuri Rasovsky, The National Radio Theater, WFMT, January 1973–April 1986).

The Mercury Theatre on the Air, "The War of the Worlds" (Orson Welles, CBS Radio, October 30, 1938).

The Fall of the City (Archibald MacLeish, Irving Reis, CBS Radio, April 11, 1937).

Neverwhere (Neil Gaiman, Heather Larmour, Dirk Maggs, BBC Radio 4, March 16, 2013–March 22, 2013, https://soundcloud.com/ger-coss-288800495/sets/neverwhere-bbc-radio).

Good Omens (Neil Gaiman, Dirk Maggs, BBC Radio 4, December 22, 2014–December 27, 2014, https://archive.org/details/good-omens-ep-6).

Anansi Boys (Neil Gaiman, Dirk Maggs, BBC Radio 4, December 25, 2017–December 30, 2017, https://soundcloud.com/penguin-books/anansi-boys-a-bbc-radio-4-full-cast-dramatisation).

The Archers (Godfrey Baseley, Julie Beckett, BBC Home Radio 4, January 1, 1951–present, https://www.bbc.co.uk/programmes/b006qpgr/episodes/player).

To Have Done with the Judgement of God (Antonin Artaud, French Radio, February 2, 1948 [scheduled, never aired], https://vimeo.com/245767392).

Sweeny Todd and the String of Pearls (Yuri Rasovsky, Hollywood Theater of the Ear, BlackStone Audio, January 1, 2006, https://www.hoopladigital.com/title/10027663).

The Maltese Falcon (Yuri Rasovsky, Hollywood Theater of the Ear, BlackStone Audio, November 2008, https://www.booksamillion.com/p/Maltese-Falcon/Dashiell-Hammett/9781433252501).

The Cabinet of Dr. Caligari (Yuri Rasovsky, Hollywood Theater of the Ear, January 1, 1998, https://www.amazon.com/The-Cabinet-of-Dr-Caligari/dp/B0007OB5VC/ref=tmm_aud_swatch_0?_encoding=UTF8&qid=&sr).

Earplay (Karl Schmidt, NPR, 1972–1990s, https://www.oldtimeradiodownloads.com/drama/earplay).

Nightfall (Bill Howell, CBC, July 4, 1980–May 13, 1983, https://podcasts.apple.com/us/podcast/nightfall/id1642609199).

The Shadow (Walter B. Gibson, Street & Smith, Mutual Broadcasting System, September 26, 1937–December 26, 1954).

Ozzie and Harriet (Ozzie Nelson, CBS, October 8, 1944–June 18, 1954).

All That Fall (Samuel Beckett, Donald McWhinnie, BBC Third Programme, January 13, 1957).

Columbia Workshop, "Ecce Homo" (Pare Lorentz, CBS, May 21, 1938).

Albert's Bridge (Charles Lefeaux, Tom Stoppard, BBC Radio 4, July 13, 1967, https://www.youtube.com/watch?v=2aCh0g-zfV8).

The Mercury Theatre on the Air, "Hell on Ice" (Orson Welles, CBS, October 9, 1938, https://orsonwelles.indiana.edu/items/show/1969).

The Dybbuk (Yuri Rasovsky, Hollywood Theater of the Ear, BlackStone Audio, April 16, 2009, https://www.kobo.com/us/en/audiobook/the-dybbuk-3).

Object Piece (Erik Bauersfeld, Drury Pifer, Randy Thom, KPFA Radio, 1976, https://www.thirdcoastfestival.org/feature/object-piece).

Dragnet (Jack Webb, NBC, June 3, 1949–July 26, 1957, https://archive.org/details/OTRR_Dragnet_Singles).
Suspense (Norman Macdonnell, William N. Robson, William Spier, Bruna Zirato Jr., CBS Radio, June 17, 1942–September 30, 1962).
Gunsmoke (Norman Macdonnell, John Meston, CBS Radio, April 26, 1952–June 11, 1961).
Dimension X (Edward King, Fred Wiehe, NBC, April 8, 1950–September 29, 1951).
Calling All Cars (James E. Davis, CBS West Coast, Mutual-Don Lee, November 29, 1933–September 8, 1939, https://archive.org/details/OTRR_Calling_All_Cars_Singles).
Mercury Summer Theater, "The Hitch-Hiker" (Lucille Fletcher, CBS Radio, November 17, 1941).
Pandemic (John Dryden, Goldhawk Productions, BBC Radio 4, April 2, 2012, https://www.audible.com/pd/Pandemic-Audiobook/B007Q409BI).
The Day That Lehman Died (John Dryden, Goldhawk Productions, BBC World Drama, September 6, 2009, https://www.bbc.co.uk/programmes/p0043kcq).
Travels with Jack (Meatball Fulton, ZBS, 1992–2000, https://www.zbs.org/index_new.php/store/jack_flanders).
Breakdown and Back (Marjorie Van Halteren, KEYA-FM, October 14, 1985–December 31, 1985).
Black Mass (Erik Bauersfeld, KPFA, 1963–67, https://www.kpfahistory.info/black_mass_index.html).
Studio One (Fletcher Markle, CBS Radio, April 29, 1947–July 27, 1948).
Somewhere Out There, "Phone Therapy" (Joe Frank, KCRW, 1997).
Suspense, "Sorry, Wrong Number" (Lucille Fletcher, William Spier, CBS Radio, May 25, 1943).
The Hitchhiker's Guide to the Galaxy (Douglas Adams, BBC Radio 4, March 8, 1978–April 12, 2018, https://www.bbc.co.uk/programmes/b03v379k/episodes/guide).
Lights Out! (Wyllis Cooper, Arch Oboler, NBC, January 3, 1934–August 6, 1947).
Quiet, Please (Wyllis Cooper, Mutual Broadcasting System, June 8, 1947–June 25, 1949).
I Love a Mystery (Carlon E. Morse, NBC West Coast, January 16, 1939–December 29, 1944).
The Fourth Tower of Inverness (Thomas Lopez [as Meatball Fulton], ZBS, 1972).
I Was a Communist for the FBI (Matt Cvetic, Frederick W. Ziv Company, April 23, 1952–October 14, 1953).
Counterspy (Phillips H. Lord, NBC Blue Network, May 18, 1942–November 29, 1957).
The Dog It Was That Died (Tom Stoppard, BBC Radio 3, December 9, 1982).
The Spy Who Came in from the Cold (John Le Carre, Patrick Rayner, BBC Radio 4, July 5, 2009).
Ballad for Americans (Norman Corwin, Paul Robeson, CBS, November 5, 1939).
Suspense, "The Black Curtain" (William Spier, Cornell Woolrich, CBS Radio, November 30, 1944).

Suspense, "Mission Completed" (John Forrest, William Spier, CBS Radio, December 1, 1949).
Suspense, "The Search for Henri Lefevre" (Lucille Fletcher, William Spier, CBS Radio, July 6, 1944).
A Comedy of Danger (Richard Hughes, BBC, January 15, 1924).
Escape, "Three Skeleton Key" (William N. Robson, CBS, November 15, 1949).

CODA

Where Is Radio?

In May 2020, a moment during the COVID pandemic in which many were just coming to recognize that the lockdown may not end up being as brief as they had initially hoped, Chicago's Experimental Sound Studio transmitted a series of broadcasts both on linear radio and online that they called Quarantine Concerts. The first was a performance by Paris-based artists DinahBird and Jean-Philippe Renoult titled "The Gold Line."[1] The transmission was also the ninety-second broadcast of the local experimental radio arts organization Radius, then celebrating its ten-year anniversary with a show called DECADE that reflected on time. Headed by Chicago artist Jeff Kolar, Radius employed an autonomous low-watt FM transmitter to broadcast a variety of works that exposed, poked fun at, and grappled with radio itself as a plastic space for thought, art, and play. To that date, Radius had broadcast work by more than two hundred artists from twenty countries, and its material has reflected the tangled roots of radio art's traditions, from avant-garde writers and revolutionaries of the early twentieth century to Free Radio activists from Italy to Japan in the 1970s and 1980s.[2] The result ranged from musical pieces based on recordings of static to reflections on the earliest broadcasts of the human voice, the creation of temporary micro-stations, broadcasts of sounds from underwater in Lake Michigan, and sonifications of the electromagnetic fields produced by power stations, microwave ovens, and baby monitors.

Establishing the parameters of "radio art" has been a subject of much debate, although artists have engaged with the medium only sporadically. Gregory Whitehead, among the most prolific creators in the genre, once remarked that the history of radio art is in a very literal sense "a history of

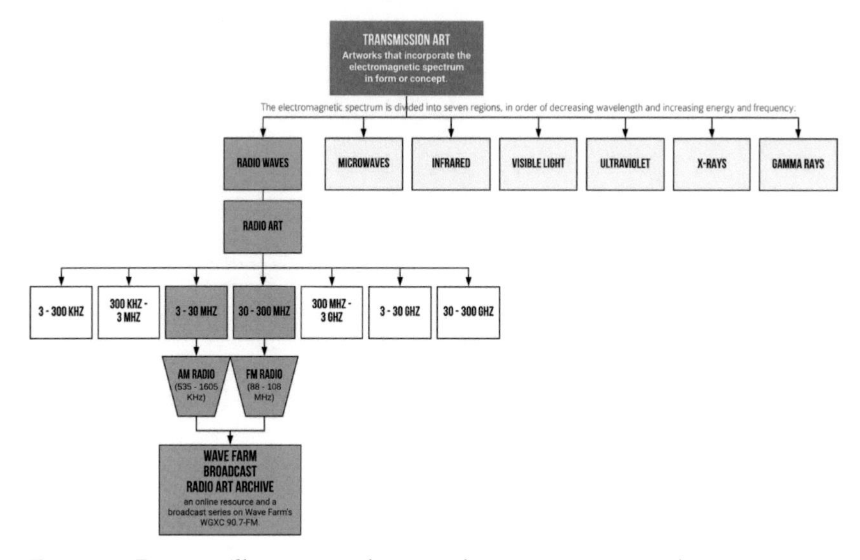

Figure 14. Diagram illustrating radio art and transmission art on the electromagnetic spectrum. Copyright Galen Joseph-Hunter. Image courtesy of Galen Joseph-Hunter and Wave Farm (Acra, NY).

nobodies," with the artist a rare and typically forgotten visitor to a medium that has historically been controlled by entrenched commercial, state, and military interests.[3] Today, many leading practitioners often distinguish between art created for radio (a concert, for instance) and more direct "radio art" in which electromagnetic waves themselves are at the center of the work as such (see fig. 14). Other approaches to defining radio art are more specific, distinguishing radio art as works that focus on the radio band (AM and FM, along with adjacent voice-based transmission frequencies) as opposed to transmission art that focuses on other parts of the spectrum (using VLF or ELF frequency ranges, for example, or moving up to the gigahertz area of microwave and beyond), although, in practice, organizations like Radius tended to feature both.[4] Kolar's frequent collaborator Anna Friz, a leading radio artist and theorist, highlights in much of her writing that, whether it takes the form of a micro-broadcast, transcontinental transmission, performance, or installation, radio art has historically exhibited a few long-standing interests. These have included appropriating forms and devices of normative broadcasting, disturbing one-to-many structures in favor of many-to-many or few-to-few, and using the material aspects of the electromagnetic spectrum to create relationships between people, devices, and objects, from cell phones and radio systems to electrical grids. For Friz, radio art gives us a way to come to grips with what she calls "transmission ecologies," the material and symbolic systems of the physical and electromagnetic environments in which we live, by intervening in "radio space," understood to be "a continuous, available, fluctuating area described by the reach of signals with overlapping fields of influence and the space of imagination that invisible territory enables."[5]

"The Gold Line" puts its finger on just such an ecology, one that was right under the noses of many who lived beneath its territory but was of great importance to the ruthless efficiency of twenty-first-century financial capitalism. Bird and Renoult's piece takes its title from the nickname of a series of microwave repeater radio towers that follow a geodesic path straight from the New York Stock Exchange's data center in New Jersey to the Chicago Mercantile Exchange's center in Aurora, Illinois. Called the "Gold Line" by high-frequency traders, the system makes it possible for algorithms to execute a trade between the two exchanges at a speed of 0.0081 seconds, giving traders an edge over other methods of transmission. In 2018, Bird and Renoult had taken a twenty-four-day journey along this geodesic through parts of rural Indiana, Ohio, and Pennsylvania, where the repeater towers of the line are tucked behind barbed wire fences and down dirt roads across a rural land marked by signs of neglect and disinvestment, a different world from that of high-speed traders, whose transmissions exhibit indifference to the lives lived below. In this way, the artists chose to react to the Gold Line's extraordinary wave speed with slow time of travel, recording atmospheres and electromagnetic transmissions near the towers, many of which were strung up along old AT&T network infrastructure, part of linear radio's fixed material history (see fig. 15). Using video, photography, and sound captured by coil microphones able to "hear" algorithm trade transmissions, the web broadcast of the piece showed images of towers over sonified static, putting listeners in direct touch with the very material electromagnetic texture by which the flows of capital that structure modern economies flowed around them, invisibly. The piece displaced these normally inaudible transmissions from the microwave area of the spectrum into the radiophonic area, specifically to 87.9 FM in the West Edgewater neighborhood of Chicago, where I heard it in a radiophonic simulcast with the web stream.

"The Gold Line" is but one of a great many compelling radio artworks that emerged in these years that can be critically engaged by means of Friz's idea of transmission ecologies, with several pieces functioning as

Figure 15. Three still images from the video portion of the web stream of "The Gold Line" by DinahBird and Jean-Philippe Renoult. Images courtesy of the artists.

ways of either identifying or creating such ecologies. For his 2014 piece *Crazy Horse One-Eight*, for example, Gregory Whitehead created a four-part cantata that drew on leaked war broadcasts involving the murder of civilians by US forces, specifically focusing on the weaponization of radiophonic space as superimposed on the sovereign lands of others. In four cantos, Whitehead sings fragments of radio conversations of soldiers about a massacre, his voice broken up by throbbing drones and muttering radiophonic detritus, reflecting the humdrum radiophony of civilian death. In 2019's *Ghosts in the Air Glow*, meanwhile, Amanda Dawn Christie employed a disused array of transmission towers in Alaska (the High-Frequency Active Auroral Research Program) to distribute sounds and images around the world, mingling her unique composition with the electromagnetic energy of the northern aurora, through the ionosphere at the edge of outer space. These transmissions could be decoded into images by shortwave listeners around the world, who sent her back old-fashioned QSL cards to mark the contact.

These works reflecting the transmission ecologies of technoscience and war could also be used for narrative work. In *The Joy Channel*, the final version of a multiyear metanarrative piece that Friz herself developed with Emmanuel Madan from 2007 to 2017, she created a complex narrative of a near future (or, perhaps more accurately, a recent present) that is revealed through stray transmissions, exploring a world in which corporate devices broadcast emotions (the titular "Joy Channel") directly to the mind, while a community of tele-empaths begin to connect through radio broadcasts and HAM radio, a salvaged technology for a time of climate change scarcity. Another innovative approach to narrative came from the world of music, in the work of electroacoustic composer Yvette Janine Jackson, whose practice of creating "radio operas" draws on the ideas of radio dramaturgy from the mid-twentieth century, bringing them to bear on large-scale works on topics including her record *Freedom*, which included one side that used transmission and found sound to explore the Middle Passage and another that explored responses to marriage equality among Black community leaders. *The Guardian* described the first as "a harrowing babble of ocean sounds, heartbeats, distorted screams, Bernard Herrmann strings and slow-motion explosions, which seems to obey an almost symphonic structure" and the second as "a restless, cut-and-clip montage of (frequently shocking) quotes from street preachers, politicians, TV evangelists and excerpts from essays by homophobic Afrocentric academics."[6] Both compositions are, like all of Jackson's radio operas, intensely radiophonic in how they invite imaginary relationships; they do not just use transmission ecologies as raw material, but they also become such ecologies, linking past and present, inside and outside.

None of these pieces are, by any reasonable definition, podcasts. That is just why I describe them here, to raise two concluding points of criticism, one of this project itself and the other of the industry that it describes. The first is that when it comes to audio, although the mid-2010s to COVID will undoubtedly be remembered as a period whose main story is the rise of narrative podcasting—this book is actively generating that story at this very moment—these same years saw a remarkable revival of radio art that went largely unnoticed outside the audio world, drowned out by the media din around podcasting. And this revival took place despite the fact that (or perhaps because) many institutions and creators were in these years abandoning the electromagnetic spectrum in favor of internet-based distribution, as national broadcasters began to draw down and even eliminate their use of traditional linear broadcast systems, while federal regulators repurposed the once publicly owned part of the spectrum for use by financial systems, satellites, private cell phone companies, broadband, or military applications—ironically, just the sorts of uses that transmission ecology radio art prompts us to recognize and to intervene in.

While this book has explored the work of audio creators whose main conceptual and artistic coordinates were *Radiolab* and *This American Life*, it has largely ignored those whose coordinates were Bertolt Brecht and Velimir Khlebnikov, John Cage and Sun Ra, Helen Thorington and Marjorie Van Halteren, Antonin Artaud and Tetsuo Kogawa, Jacki Apple and Aki Onda.[7] While I have been describing meetings like HearSay, Podcast

Movement, and the Third Coast Festival, along with the online exchanges among their communities, I have not similarly approached the locations where radio art grew and prospered, from the Deep Wireless Festival in Toronto to the Radio Revolten monthlong radio art festival in Germany and several Radiophrenia festivals from the Glasgow Center for Contemporary Arts featuring up to two weeks of radio artwork, twenty-four hours a day. While these pages have focused on come-and-go podcast start-ups in Brooklyn flush with venture capital, it has ignored La Radia, a scrappy network of radio art stations from London's Resonance FM to Halle's Radio Corax, San Marino's Usmaradio, Dunedin's Radio One, Montréal's CKUT, and Vienna's Kunstradio. At New York's Wave Farm, another of these stations, a number of radio fellows and artists and archivists worked in these same years to build an archive of radio art, one currently housing more than seven hundred curated pieces.[8] If the story of the rise of narrative podcasting has an unconscious, then it is surely these moments and places, where radio was being rethought, reimagined, and reinvented with as much energy and ambition as at any time in history. Indeed, the issues that radio artists described above engaged with so powerfully—racial inequality, the climate crisis, imperial war, marriage equality, hypercapitalism—are very much the issues that narrative podcasters hoped to speak to as well, and the frameworks and concepts that radio art brought to these topics achieved levels of insight with them that serialized narrations could not.

What made radio art such a compelling approach is precisely, and this is my second point, the fact that focus on transmission ecology rooted radio art in place. That is, even though radio is now, and has always been, a medium based on the principle of dissemination, of speaking into a void without really knowing who is out there (Whitehead is right when he describes radio as a "ghostland/dreamland"), it is nevertheless the case that broadcasts know where they come from and tell us so. For a hundred years the medium has grown poetically structured around the idea that everything it did had an origin point, a call sign, a "station," a place at the center of the fluctuating space that the signal reaches, where the transmission believes it belongs. This is just what is missing from podcasting. In pursuing what felt like limitless distribution, podcasts lost a sense of being *from* somewhere, inside a particular ecology that can be investigated, named, revealed, and contested. We get our podcasts from the internet, a space that is composed of many objects and territories but that still feels like we're getting them from nowhere, and podcasting has done little to disabuse its listeners of that illusion by calling attention to the medium's basic conditions of possibility. Indeed, many works of audio narrative were more thematically fascinated by analog tape than they were by their own digital architectures, a self-indifference that radio artists consistently refused to indulge.

In the late 2010s, as I have argued, much of the thought and practice in podcasting came from those captivated by its definition, "What is a podcast?"—a question around whose lack of resolution flowed a mysterious generative energy that wrapped virtually all the pieces described in this book in a misty glow through which rolled new ways of mapping obsession, fascinations with epistemology, and an aestheticization of amnesia. This glow obscured for a moment the equally crucial philosophical question of *where* a podcast is, materially and also conceptually, and how we would engage with its unique transmission ecology, exactly the topic on which many radio artists ruminated in these same years with such ingenuity. In comparison to radio art, which was compulsively asking and answering the question of "where" again and again, podcast pieces always feel a little homeless, their material support distanced from the text, mystified by the sheer forwardness of propulsive elements like story, character, and plot—of narrativity itself. That left the question of materiality oddly muted. As the generative glow of the narrative era fades, one thing that podcasting as a concept and practice clearly still has to learn, and its radiophonic unconscious may be the best teacher, is the art of how to be someplace in the material world. There is still a long way to go.

Radio Works Discussed in Order of Appearance in the Text

DinahBird and Jean-Philippe Renoult, "The Gold Line" (May 15, 2020, Radius and the Experimental Sound Studio, Chicago, https://bird-renoult.net/the-gold-line/).

Gregory Whitehead, *Crazy Horse One-Eight* (Commissioned for Radio Dreamlands project, Radio Arts UK, May 30, 2014, https://gregorywhitehead.net/2014/05/30/crazy-horse-one-eight/).

Amanda Dawn Christie, *Ghosts in the Air Glow* (Commissioned by the Canada Council for the Arts, transmitted from the Anchorage Radio Science Operations Centre, Anchorage, Alaska, March 25, 2019, https://ghostsintheairglow.space/archived-videos-2019).

Anna Friz and Emmanuel Madan, *The Joy Channel* (IO Sound, September 4, 2018, https://iosound.ca/2018/09/iota-001/).

Yvette Janine Jackson, *Freedom* (Fridman Gallery, July 25, 2021, https://yvettejaninejackson.bandcamp.com/album/freedom).

Notes

Introduction

1. Becca James, "Podcasts Are Always the Next Big Thing," *Vulture*, September 4, 2020, https://www.vulture.com/article/timeline-podcasts-news-media-stories.html; Ben Hammersley, "Audible Revolution," *The Guardian*, February 11, 2004, https://www.theguardian.com/media/2004/feb/12/broadcasting.digitalmedia.

2. André Bazin, *What Is Cinema?* Vols. 1 and 2 (Berkeley: University of California Press, 2004).

3. On this change, see Nicholas Quah, "Apple's Podcast Analytics Reckoning Is Here," *Vulture*, December 15, 2017, https://www.vulture.com/2017/12/apple-podcast-analytics-listener-data.html; Sarah Perez, "Apple Launches Its Podcast Analytics Service into Beta," Techcrunch, December 14, 2017, https://techcrunch.com/2017/12/14/apple-launches-its-podcast-analytics-service-into-beta/.

4. On advertising revenue, listenership numbers, and the number of podcasts, see, in order of mention, Jonathan Berr, "Podcasting Is Finally Attracting Real Money," CBS News, September 14, 2015, https://www.cbsnews.com/news/podcasting-is-finally-attracting-real-money/; "The Infinite Dial 2015," Edison Research, last modified March 4, 2015, http://www.edisonresearch.com/wp-content/uploads/2015/03/InfiniteDial2015.pdf; "The Infinite Dial 2019," Edison Research, last modified March 6, 2019, https://www.slideshare.net/webby2001/infinite-dial-2019; Maya Dukmasova, "Radio Gaga: An Ear to the Ground at 'The Sundance of Radio' in the Golden Age of Podcasting," *Chicago Reader*, December 14, 2017, 14; Ben Sisario, "Podcasting Is Booming. Will Hollywood Help or Hurt Its Future," *New York Times*, February 25, 2021, https://www.nytimes.com/2021/02/25/arts/podcasts-hollywood-future.html; "Podcast Stats: How Many Podcasts Are There?," Listen Notes, accessed February 8, 2023, https://www.listennotes.com/podcast-stats/.

5. Mason Walker, "Nearly a Quarter of Americans Get News from Podcasts," Pew Research Center, February 15, 2022, https://www.pewresearch.org/fact-tank/2022/02/15/nearly-a-quarter-of-americans-get-news-from-podcasts/.

6. This has always been true of radio relative to its competing media. Although narratives about the rise of TV in the 1950s typically suggest that the medium ended the radio age, many scholars have shown radio growing and evolving to this day. See, e.g., Susan Douglas, *Listening In: Radio and the American Imagination* (Minneapolis: University of Minnesota Press, 2013), 220.

7. "State of the Media: Audio Today, How America Listens," Nielsen Company, last modified June 23, 2017, https://www.nielsen.com/insights/2017/state-of-the-media-audio-today-2017/.

8. "PwC's Annual Report Points to More Growth for Radio and Other Audio Industries," Inside Radio, June 22, 2022, https://www.insideradio.com/free/pwc-s-annual-report-points-to-more-growth-for-radio-and-other-audio-industries/article_17bb6b1e-f204-11ec-a093-f38366aed18f.html.

9. Dolores Inés Casillas, *Sounds of Belonging: US Spanish-Language Radio and Public Advocacy* (New York: NYU Press, 2014), 7. We are blessed to live in a golden age of texts on Spanish language radio in the United States. See, e.g., Christopher Chávez, *The Sound of Exclusion: NPR and the Latinx Public* (Phoenix: University of Arizona Press, 2021); Monica De La Torre, *Feminista Frequencies: Community Building through Radio in the Yakima Valley* (Seattle: University of Washington Press, 2022); Sonia Robles, *Mexican Waves: Radio Broadcasting along Mexico's Northern Border, 1930–1950* (Phoenix: University of Arizona Press, 2019).

10. Bello Collective, "Sometimes, We Created the Conversation: A Look Back at Bell's Contributions to the Industry," Medium, January 12, 2022, https://bellocollective.com/we-created-the-conversation-2597002a6d98.

11. Galen Beebe, "You Think There's No Podcast Criticism? Think Again," Medium, September 24, 2020, https://bellocollective.com/you-think-theres-no-podcast-criticism-think-again-360c89ee775b; Erik Jones, "The Thriving Ecosystem of Podcast Media," Medium, March 16, 2021, https://bellocollective.com/the-thriving-ecosystem-of-podcast-media-13adf43e0aee.

12. See, e.g., Carolyn Marvin, *When Old Technologies Were New: Thinking about Electric Communication in the Late Nineteenth Century* (New York: Oxford University Press, 1988); Matthew C. Ehrlich, *Radio Utopia: Postwar Audio Documentary in the Public Interest* (Champagne-Urbana: University of Illinois Press, 2011); Michele Hilmes, *Radio Voices: American Broadcasting, 1922–1952* (Minneapolis: University of Minnesota Press, 1997); Wendy Hui Kyong Chun, *Updating to Remain the Same: Habitual New Media* (Boston: MIT Press, 2016); Paul Starr, *The Creation of the Media: Political Origins of Modern Communications* (New York: Basic Books, 2004).

13. Lisa Gitelman, *Always Already New: Media, History, and the Data of Culture* (Boston: MIT Press, 2008), 6.

14. For these background texts, see Neil Verma, "From the Narrator's Lips to Yours: Streaming, Podcasting, and the Risqué Aesthetic of Amazon Channels," *Participations* 16, no. 2 (2019): 273–97; Neil Verma, "Recessive Epistemologies in True Crime Podcasting," in *The Routledge Companion to Radio and Podcast Studies*, ed. Mia Lindgren and Jason Loviglio (New York: Routledge, 2022), 179–89; Neil Verma, "The Arts of Amnesia: The Case for Audio Drama," *RadioDoc Review* 3, no. 1 (2017); Neil Verma, "*Serial* Goes Missing," Antenna, January 8, 2016, https://blog.commarts.wisc.edu/2016/01/08/serial-goes-missing/; Neil Verma, "Pillow, Talk," *RadioDoc Review* 4, no. 1 (2018).

15. Lucas Shaw, "Podcasting Hasn't Produced a New Hit in Years," *Bloomberg*, January 9, 2022, https://www.bloomberg.com/news/newsletters/2022-01-09/podcasting-hasn-t-produced-a-new-hit-in-years.

16. "Podcasting Statistics: Data from Study of 1,076 Podcasters," Improve Podcast, https://improvepodcast.com/podcasting-statistics-data/.

17. Sisario, "Podcasting Is Booming."

18. Peter Allen Clark, "Podcasts Lose Their Edge" Axios, February 9, 2023, https://www.axios.com/2023/02/09/podcasts-lose-edge-spotify.

19. "Podcast Stats: How Many Podcasts Are There?," Listen Notes, accessed February 8, 2023, https://www.listennotes.com/podcast-stats/.

20. Nicholas Quah, "Podcasting Is Just Radio Now," *Vulture*, September 22, 2022, https://www.vulture.com/2022/09/podcasting-is-just-radio-now.html; Max Tani, "How Spotify's Podcast Bubble Went Wrong" Semafor, February 13, 2023, https://www.semafor.com/article/02/12/2023/how-spotifys-podcast-bet-went-wrong. Personality-driven podcasting had always been an important element of the field. On this, see Cwynar's highly convincing work: Christopher Cwynar, "Self-Service Media: Public Radio Personalities, Reality Podcasting, and Entrepreneurial Culture," *Popular Communication* 17, no. 4 (2019): 317–32.

21. See, e.g., D. N. Rodowick, *The Virtual Life of Film* (Cambridge, MA: Harvard University Press, 2007); Derek Johnson, ed., *From Networks to Netflix: A Guide to Changing Channels*, 2nd ed. (New York: Routledge, 2022); Aymar Jean Christian, *Open TV* (New York: NYU Press, 2018); Amanda D. Lotz, *The Television Will Be Revolutionized* (New York: NYU Press, 2014); Shane Denson and Julia Leyda, eds., *Post-Cinema: Theorizing 21st-Century Film* (Falmer, UK: Reframe Books, 2016), https://reframe.sussex.ac.uk/reframebooks/archive2016/post-cinema-denson-leyda/.

22. See, e.g., Siobhán McHugh, *The Power of Podcasting: Telling Stories through Sound* (Sydney: University of New South Wales Press, 2022). Two key texts capturing the invention of podcast studies are Dario Llinares, Neil Fox, and Richard Berry, *Podcasting: New Aural Cultures and Digital Media* (New York: Palgrave MacMillan, 2018); and Andrew J. Bottomley, "Podcasting: A Decade in the Life of a 'New' Audio Medium: Introduction." *Journal of Radio & Audio Media* 22, no. 2 (2015): 164–69.

23. See Vanessa Quirk, "Guide to Podcasting," *Columbia Journalism Review*, December 7, 2015, https://www.cjr.org/tow_center_reports/guide_to_podcasting.php.

24. Jeremy Wade Morris, Samuel Hansen, and Eric Hoyt, "The PodcastRE Project: Curating and Preserving Podcasts (and Their Data)," *Journal of Radio & Audio Media* 26, no. 1 (2019): 8–20.

25. Andrew Bottomley, *Sound Streams: A Cultural History of Radio-Internet Convergence* (Ann Arbor: University of Michigan Press, 2020).

26. Bottomley, *Sound Streams*, 127.

27. Jonathan Sterne, Jeremy Morris, Michael Brendan Baker, and Ariana Moscote Freire, "The Politics of Podcasting," *Fibreculture Journal*, December 13, 2008, https://fibreculturejournal.org/fcj-087-the-politics-of-podcasting/.

28. Michele Hilmes, "But Is It Radio? New Forms and Voices in the Audio Private Sphere," in *The Routledge Companion to Radio and Podcast Studies*, ed. Mia Lindgren and Jason Loviglio (New York: Routledge, 2022), 9–18.

29. Henry Jenkins, *Convergence Culture: Where Old and New Media Collide* (New York: NYU Press, 2008); Jay David Bolter and Richard Grusin, *Remediation: Understanding New Media* (Cambridge, MA: MIT Press, 1999).

30. Martin Spinelli and Lance Dann, *Podcasting: The Audio Media Revolution* (New York: Bloomsbury Academic, 2019), 7–8.

31. Rodowick, *The Virtual Life of Film*, 28.

32. Richard Berry, "What Is a Podcast? Mapping the Technical, Cultural, and Sonic Boundaries between Radio and Podcasting," in *The Routledge Companion to Radio and Podcast Studies*, ed. Mia Lindgren and Jason Loviglio (New York: Routledge, 2022), 399–407. Berry's thought on the relationship between radio and podcasting over two decades is perhaps the most sophisticated and layered of any scholar in the field. See, e.g., Richard Berry, "Will the iPod Kill the Radio Star? Profiling Podcasting as Radio," *Convergence* 12, no. 2 (2006): 143–62; Richard Berry, "A Golden Age of Podcasting? Evaluating *Serial* in the Context of Podcast Histories," *Journal of Radio & Audio Media* 22, no. 2 (2016): 170–78; and Richard Berry, "Part of the Establishment: Reflecting on 10 Years of Podcasting as an Audio Medium," *Convergence* 22, no. 6 (2016): 661–71.

33. *Oxford English Dictionary*, s.v. "podcast (noun)," accessed December 12, 2022, https://www-oed-com.turing.library.northwestern.edu/view/Entry/273002. See also: Matthew Honan, "Podcast is 2005 Word of the Year" *Macworld*, December 5, 2005, https://www.macworld.com/article/178001/podcastword.html.

34. "The Infinite Dial 2016," Edison Research, last modified March 10, 2016, https://www.edisonresearch.com/the-infinite-dial-2016/.

35. "The Podcast Consumer 2019," Edison Research, last modified April 5, 2019, https://www.slideshare.net/webby2001/edison-research-podcast-consumer-2019; "Super Listeners 2021," Edison Research, last modified February 16, 2022, https://www.edisonresearch.com/super-listeners-2021-from-edison-research-and-ad-results-media/; Natalie Jarvey, "Prince Harry, Meghan Markle Strike Spotify Podcast Deal," *Hollywood Reporter*, December 15, 2020, https://www.hollywoodreporter.com/business/digital/prince-harry-meghan-markle-strike-spotify-podcast-deal-4105096/#!.

36. Tiziano Bonini, "Podcasting as a Hybrid Cultural Form between Old and New Media," in *The Routledge Companion to Radio and Podcast Studies*, ed. Mia Lindgren and Jason Loviglio (New York: Routledge, 2022), 19–29. See also Robert S. Boynton, "Intimacy, Inc.," *RadioDoc Review* 8, no. 1 (2022); and Jemily Rime, Chris Pike, and Tom Collins, "What Is a Podcast? Considering Innovations in Podcasting through the Six-Tensions Framework," *Convergence* 28, no. 5 (June 2022): 1260–82.

37. See Kim Fox, David O. Dowling, and Kyle Miller, "A Curriculum for Blackness: Podcasts as Discursive Cultural Guides, 2010–2020," *Journal of Radio & Audio Media* 27, no. 2 (2020): 298–318; Sarah Florini, "This Week in Blackness, the George Zimmerman Acquittal, and the Production of a Networked Collective Identity," *New Media & Society* 19, no. 3 (2017): 439–54.

38. Neil Verma, *Theater of the Mind: Imagination, Aesthetics, and American Radio Drama* (Chicago: University of Chicago Press, 2012).

39. See, e.g., Jessica Abel, *Out on the Wire: The Storytelling Secrets of the New Masters of Radio* (New York: Broadway Books, 2015); John Biewen and Alexa Dilworth, eds., *Reality Radio: Telling True Stories in Sound*, 2nd ed. (Chapel Hill: University of North Carolina Press, 2017); McHugh, *The Power of Podcasting*; Leslie Grace McMurtry, *Revolution in the Echo Chamber: Audio Drama's Past, Present and Future* (Bristol, UK: Intellect Books, 2019); Sven Preger, *Storytelling in Radio and Podcasts: A Practical Guide* (New York: Palgrave MacMillan, 2021); Spinelli and Dann, *Podcasting*; Kc Wayland, *Bombs Always Beep: Creating Modern Audio Theater* (Orange, CA: Wayland Productions, 2018).

40. See https://transom.org and https://radiodramarevival.com.

41. Eric Hoyt, J. J. Bersch, Susan Noh, Samuel Hansen, Jacob Mertens, and Jeremy Wade Morris, "PodcastRE Analytics: Using RSS to Study the Cultures and Norms of Podcasting," *DHQ: Digital Humanities Quarterly* 15, no. 1 (2021).

42. See, e.g., Nicholas L. Baham III and Nolan Higdon, *The Podcaster's Dilemma: Decolonizing Podcasters in the Era of Surveillance Capitalism* (New York: John Wiley & Sons, 2022); Corrina Laughlin, "The Millennial Medium: The Interpretive Community of Early Podcast Professionals," *Television & New Media* (2023): 1–15.

43. See https://www.equalityinaudiopact.co.uk/.

44. Abel, *Out on the Wire*, 11.

45. Jonathan Kern, *Sound Reporting: The NPR Guide to Audio Journalism and Production* (Chicago: University of Chicago Press, 2008), 95.

46. Alex Blumberg, "Power Your Podcast with Storytelling," CreativeLive, 2014, https://www.creativelive.com/class/power-your-podcast-storytelling-alex-blumberg.

47. Abel, *Out on the Wire*, 115; Preger, *Storytelling in Radio and Podcasts*, 14.

48. Marie-Laure Ryan, "Radio Drama between Mimetic and Diegetic Presentation," in *Audionarratology: Lessons from Radio Drama*, ed. Lars Bernaerts and Jarmila Mildorf (Columbus: The Ohio State University Press, 2021), 215–23.

49. Alix Spiegel, "Variations in Tape Use and the Position of the Narrator: Alix Spiegel's Practical Guide to Different Radio Techniques," in *Reality Radio: Telling True Stories in Sound*, 2nd ed., ed. John Biewen and Alexa Dilworth (Chapel Hill: University of North Carolina Press, 2017), 46.

50. For the Third Coast Audio Library, see https://www.thirdcoastfestival.org/overview/library.

51. McHugh, *The Power of Podcasting*, 42–46.

52. I am very grateful to Peabody Collection archivist Mary L. Miller, who was able to help us acquire the raw metadata we needed, even in a challenging period of the pandemic.

53. For transparency, I should point out that I chaired one of several screening committees for this award during these years. However, because the metadata came from the total set of all submissions, and since my input took place subsequent to the submission process, my screening role had no direct effect on the composition of these data.

Chapter 1

1. Cartwright, "How People Obsess over 'Serial': Funny or Die," YouTube, 2:57, November 13, 2014. https://www.youtube.com/watch?v=AZqUyPT_Xik.

2. *Serial*, "Oh My God, We're Obsessed," Facebook, November 4, 2014, https://www.facebook.com/serialpodcast/posts/782978068428576.

3. Andrew Norton, "This Is Radio: Five Questions," Transom, video, 6:36, December 11, 2014 https://transom.org/2014/this-is-radio-5-questions.

4. Vanessa Quirk, "Guide to Podcasting," *Columbia Journalism Review* 7 (December 7, 2015); Siobhán McHugh, "Michelle Obama, Podcast Host: How Podcasting Became a Multi-Billion Dollar Industry," *The Conversation*, July 28, 2020, https://theconversation.com/michelle-obama-podcast-host-how-podcasting-became-a-multi-billion-dollar-industry-142920.

5. Nicholas Quah, "Apple's Podcast Analytics Reckoning Is Here," *Vulture*, December 15, 2017, https://www.vulture.com/2017/12/apple-podcast-analytics-listener-data.html.

6. Seth Resler, "Understanding Podcast Stats: Q&A with Libsyn's VP of Podcaster Relations Rob Walch," All Access, September 30, 2015, https://www.allaccess.com/podcast/broadcaster-meets-podcaster/archive/22943/understanding-podcast-stats-q-a-with-libsyn-s-vp; "The Infinite Dial 2015," Edison Research, March 4, 2015, https://www.edisonresearch.com/the-infinite-dial-2015.

7. Erica Haugtvedt, "The Ethics of Serialized True Crime: Fictionality in *Serial* Season One," in *The "Serial" Podcast and Storytelling in the Digital Age*, ed. Ellen McCracken (New York: Routledge, 2017), 18.

8. Nicholas Quah, "Hot Pod: Is 'Why Doesn't Audio Go Viral?' the Wrong Question to Ask?," Nieman Journalism Lab, December 1, 2015, niemanlab.org.

9. Martin Spinelli and Lance Dann, *Podcasting: The Audio Media Revolution* (New York: Bloomsbury, 2019), 48–60; Sarah Florini, *Beyond Hashtags: Racial Politics and Black Digital Networks* (New York: NYU Press, 2019).

10. Siobhán McHugh, *The Power of Podcasting: Telling Stories through Sound* (Sydney: University of New South Wales Press, 2022), 113–15.

11. Ernesto Londoño, "Hooked on the Freewheeling Podcast 'Serial,'" *New York Times*, February 12, 2015; Terry Gross, "Serial Host Sarah Koenig Says She Set Out to Report, Not Exonerate," NPR, *Fresh Air* transcript, December 23, 2014, https://www.npr.org/transcripts/372577482.

12. Miranda Sawyer, "*Serial* Review—The Greatest Murder Mystery You Will Ever Hear," *The Guardian*, November 8, 2014, https://www.theguardian.com/media/2014/nov/08/serial-review-greatest-murder-mystery-ever-hear.

13. Dan Fitchette, "5 Reasons Everyone's Obsessed with 'Serial,'" *Vulture*, November 6, 2014, https://www.vulture.com/2014/11/serial-podcast-why-is-everyone-obsessed.html; James Wolcott, "So, Like, Why Are We So Obsessed with Podcasts Right Now?" *Vanity Fair*, January 8, 2016, https://www.vanityfair.com/culture/2016/01/james-wolcott-on-podcasts.

14. Allie Volpe, "Your True Crime Obsession Could Be Hurting Your Mental Health," *Vice*, April 12, 2021, https://www.vice.com/en/article/v7e4b9/true-crime-and-mental-health.

15. Nancy Updike, "Better Writing through Radio," Transom, January 24, 2006, https://transom.org/2006/nancy-updike.

16. Jessica Abel, *Out on the Wire: The Storytelling Secrets of the New Masters of Radio* (New York: Broadway Books, 2015), 32–56.

17. Sianne Ngai, *Our Aesthetic Categories: Zany, Cute, Interesting* (Cambridge, MA: Harvard University Press, 2012), 112.

18. Ngai, *Our Aesthetic Categories*, 132–33.

19. See Giuseppe Craparo, Francesca Ortu, and Onno van der Hart, eds., *Rediscovering Pierre Janet: Trauma, Dissociation, and a New Context for Psychoanalysis* (New York: Routledge, 2019).

20. Jan Goldstein, *Console and Classify: The French Psychiatric Profession in the Nineteenth Century*, 2nd ed. (Chicago: University of Chicago Press, 2001), 153.

21. Marina van Zuylen, *Monomania: The Flight from Everyday Life in Literature and Art* (Ithaca: Cornell University Press, 2018), 193.

22. Nicholas Quah, "P.J. Vogt on *Reply All*'s Instantly Legendary Episode," *Vulture*, March 11, 2020, https://www.vulture.com/2020/03/reply-all-case-of-missing-hit-interview.html; Hannah J. Davies, "Reply All's The Case of the Missing Hit: Could This Be the Best Podcast Episode Ever?" *The Guardian*, March 10, 2020, https://www.theguardian.com/tv-and-radio/2020/mar/10/reply-alls-the-case-of-the-missing-hit-could-this-be-the-best-podcast-episode-ever. I am grateful to Jacob Smith for reminding me of this episode.

23. Lauren Kranc, "The One Podcast Episode That Makes Everything Better for At Least an Hour," *Esquire*, March 20, 2020, https://www.esquire.com/entertainment/music/a31789727/reply-all-podcast-song-episode-the-case-of-the-missing-hit/.

24. Michael Fried, *Absorption and Theatricality: Painting and Beholder in the Age of Diderot* (Chicago: University of Chicago Press, 1980), 7.

25. On humorous silences in radio, see Andrew Crisell, *Understanding Radio* (New York: Routledge, 2006), 53.

26. See Jason Mittell, "Downloading *Serial* (Part 1)," Antenna, October 13, 2014, https://blog.commarts.wisc.edu/2014/10/13/downloading-serial-part-1/; Jay Caspian Kang, "White Reporter Privilege," The Awl, November 12, 2014, https://www.theawl.com/2014/11/white-reporter-privilege.

27. Vladimir Propp, *Morphology of the Folktale*, trans. Laurence Scott (Austin: University of Texas Press, 1968), 21. See also Gerald Prince, *Dictionary of Narratology*, rev. ed. (Lincoln: University of Nebraska Press, 2003), 36–37.

28. Neil Verma, *Theater of the Mind: Imagination, Aesthetics, and American Radio Drama* (Chicago: University of Chicago Press, 2012).

29. Goldstein, *Console and Classify*, 171.

30. Dorrit Cohn, "Transparent Minds: Narrative Modes for Presenting Consciousness in Fiction," excerpted in *Theory of the Novel: A Historical Approach*, ed. Michael McKeon (Baltimore: Johns Hopkins University Press, 2000), 504.

31. Alexander Russo, "Wondering about *Radiolab*: The Contradictory Legacy of Corwin in Contemporary 'Screen Radio,'" in *Anatomy of Sound: Norman Corwin and Media Authorship*, ed. Jacob Smith and Neil Verma (Oakland: University of California Press, 2016), 233–51.

32. Nicholas Quah, "Mogul: The Life and Death of Chris Lighty Is a Fascinating New Podcast about Mental Health and the Music Industry," *Vulture*, June 30, 2017, https://www.vulture.com/2017/06/podcast-review-mogul-the-life-and-death-of-chris-lighty.html.

33. Michele Hilmes, "But Is It Radio? New Forms and Voices in the Audio Private Sphere," in *The Routledge Companion to Radio and Podcast Studies*, ed. Mia Lindgren and Jason Loviglio (London: Routledge, 2022), 12.

34. Dan Misener, "The Podcast Ecosystem Is Comprised of Distinct 'Neighbourhoods,'" *Pacific Content*, May 14, 2020, https://blog.pacific-content.com/the-podcast-ecosystem-is-made-up-of-distinct-neighborhoods-9e4ec105026e.

35. David Folkenflik, "'New York Times' Retracts Core of Hit Podcast Series 'Caliphate' on ISIS," NPR, December 18, 2020, https://www.npr.org/2020/12/18/944594193/new-york-times-retracts-hit-podcast-series-caliphate-on-isis-executioner.

36. Sasha Lekach, "Here's Why Everyone's Obsessed With That *Missing Richard Simmons* Podcast," Mashable, March 20, 2017, https://test.mashable.com/article/missing-richard-simmons-podcast-popular-addictive.

37. I have always liked that Rick Altman called the high direct / reflection ratio a way of giving the listener a feeling of "for-me-ness." Rick Altman, "Afterword: A Baker's Dozen of Terms for Sound Analysis," in *Sound Theory Sound Practice*, ed. Rick Altman (New York: Routledge, 1992), 250.

38. I have also made a different sort of argument about intimacy in the context of 1930s radio drama aesthetics. See Verma, *Theater of the Mind*, 57–72.

39. Mack Hagood, *Hush: Media and Sonic Self-Control* (Durham, NC: Duke University Press, 2019).

40. "The Infinite Dial 2015," Edison Research, last modified March 4, 2015, http://www.edisonresearch.com/wp-content/uploads/2015/03/InfiniteDial2015.pdf.

41. René Girard, *A Theatre of Envy: William Shakespeare* (New York: Oxford University Press, 1991), 9.

42. Nicholas L. Baham III and Nolan Higdon, *The Podcaster's Dilemma: Decolonizing Podcasters in the Era of Surveillance Capitalism* (New York: John Wiley & Sons, 2022), 18–21, 71.

43. See Nina Sun Eidsheim, *The Race of Sound: Listening, Timbre, and Vocality in African American Music* (Durham, NC: Duke University Press, 2019), 11–12.

44. For MacArthur's test cases and a deeper explanation of these tools, see Marit J. MacArthur, Georgia Zellou, and Lee M. Miller, "Beyond Poet Voice: Sampling the (Non-)Performance Styles of 100 American Poets," *Journal of Cultural Analytics* 3, no. 1 (2018). Currently, the tools are housed at the SpokenWeb project (https://drift4.spokenweb.ca/) and the source code is available on GitHub (https://github.com/sarahayu/drift4).

45. Martha Feldman, "Voice Gap Crack Break," in *The Voice as Something More*, ed. Martha Feldman and Judith Zeitlin (Chicago: University of Chicago Press, 2019), 191; Mladen Dolar, *A Voice and Nothing More* (Boston: MIT Press, 2006), 15.

46. Roland Barthes, "The Grain of the Voice," in *Image-Music-Text*, trans. Stephen Heath (New York: Noonday Press, 1977), 188; see also Spinelli and Dann, *Podcasting*, 190.

47. Michel Chion, *Sound: An Acoulogical Treatise*, trans. James A. Steintrager (Durham, NC: Duke University Press, 2016), 103. See also Bill Kirkpatrick, "Voices Made for Print: Crip Voices on the Radio," in *Radio's New Wave*, ed. Jason Loviglio and Michele Hilmes (New York: Routledge, 2013), 116–35.

48. Goldstein, *Console and Classify*, 198.

49. See Virginia Madsen, "Your Ears Are a Portal to Another World: The New Radio Documentary Imagination and the Digital Domain," in *Radio's New Wave*, ed. Jason Loviglio and Michele Hilmes (New York: Routledge, 2013), 136–54. See also Jacob Smith and Neil Verma, eds., *Anatomy of Sound: Norman Corwin and Media Authorship* (Oakland: University of California Press, 2016).

50. Mieke Bal, *Narratology: Introduction to the Theory of Narrative*, 2nd ed. (Toronto: University of Toronto Press, 1997), 101–3. See also Gérard Genette, *Narrative Discourse: An Essay in Method*, trans. Jane E. Lewin (Ithaca: Cornell University Press, 1980), 86–112.

51. Eric Hoyt, J. J. Bersch, Susan Noh, Samuel Hansen, Jacob Mertens, and Jeremy Wade Morris, "PodcastRE Analytics: Using RSS to Study the Cultures and Norms of Podcasting," *DHQ: Digital Humanities Quarterly* 15, no. 1 (2021).

52. Sarah Larson, "'Slow Radio,' The Podcast That Promotes Monks, Moose, and Inner Peace," *New Yorker*, January 4, 2019, https://www.newyorker.com/culture/podcast-dept/slow-radio-the-podcast-that-promotes-monks-moose-and-inner-peace.

53. Ariana Martinez, "A Time-Based Medium," Transom, updated September 6, 2022, https://transom.org/2022/ariana-martinez-part-1/.

Chapter 2

1. "Previous Programs," HearSay Audio Arts Festival, accessed January 20, 2023, https://hearsayhomefires.ie/wp-content/uploads/2022/06/HS_20192021_programmes_low.pdf.

2. In these years, John Hockenberry, Garrison Keillor, and Leonard Lopate, among other long-standing public radio hosts, left their posts following news stories of impropriety, while podcast hosts like *Sword and Scale*'s Mike Boudet lost their roles after making misogynistic comments. In time, other leading figures in podcasting, such as Andy Mills of the *New York Times* and P. J. Vogt of *Reply All*, would also move on from their roles, following calls for their resignations over concerns about problematic workplace behavior.

3. Phoebe Wang, "The 2018 Third Coast Awards Ceremony Acceptance Speeches," Medium, October 8, 2018, https://medium.com/@ThirdCoastFest/the-2018-third-coast-awards-ceremony-acceptance-speeches-featuring-a-transcript-of-phoebe-wangs-e22f293e73a7.

4. On this terminology, see Karma R. Chávez, "Understanding Migrant Caravans from the Place of Place Privilege," *Departures in Critical Qualitative Research* 8, no. 1 (2019): 9–16.

5. Gérard Genette, *Narrative Discourse: An Essay in Method*, trans. Jane E. Lewin (Ithaca: Cornell University Press, 1980), 198.

6. Matthias Steup and Ram Neta, "Epistemology," in *The Stanford Encyclopedia*

of Philosophy (Fall 2020 Edition), ed. Edward N. Zalta, Stanford University, https://plato.stanford.edu/archives/fall2020/entries/epistemology.

7. Michelle Macklem, “Empathy, Ethics and Aesthetics in *Love + Radio*,” *RadioDoc Review* 3, no. 1 (2017), https://ro.uow.edu.au/rdr/vol3/iss1/4/.

8. “Library of Congress Acquires Kitchen Sisters’ Audio Archive,” Library of Congress, updated January 10, 2022, https://newsroom.loc.gov/news/library-of-congress-acquires-kitchen-sisters—audio-archive/s/ddb33b8d-2b02–4c86–9a68–0cd8ca83c34c.

9. Diana Taylor, *The Archive and the Repertoire: Performing Cultural Memory in the Americas* (Durham: Duke University Press, 2003). My thinking on archontic media production in a digital space is also shaped by Abigail De Kosnik, *Rogue Archives: Digital Cultural Memory and Media Fandom* (Cambridge, MA: MIT Press, 2016).

10. The rise of history podcasts was, of course, recognized by scholars at the time. See, e.g., Andrew J. Salvati, “Podcasting the Past: *Hardcore History*, Fandom, and DIY Histories,” *Journal of Radio & Audio Media* 22, no. 2 (2015): 231–39. On Longworth’s approach to archival voices of women in particular, see Jennifer O’Meara, *Women’s Voices in Digital Media: The Sonic Screen from Film to Memes* (Austin: University of Texas Press, 2022), 119–45.

11. Alix Spiegel, “Variations in Tape Use and the Position of the Narrator: Alix Spiegel’s Practical Guide to Different Radio Techniques,” in *Reality Radio: Telling True Stories in Sound*, 2nd ed., ed. John Biewen and Alexa Dilworth (Chapel Hill: University of North Carolina Press, 2017), 42–53.

12. For the key source texts, see Raymond Williams, “From Preface to Film (UK 1954),” in *Film Manifestos and Global Cinema Cultures: A Critical Anthology*, ed. Scott MacKenzie (Oakland: University of California Press, 2020), 607–13; Raymond Williams, *The Long Revolution* (New York: Columbia University Press, 1961); Raymond Williams, *Marxism and Literature* (Oxford: Oxford University Press, 1977).

13. Williams, *Marxism and Literature*, 134.

14. Stuart Middleton, “Raymond Williams’s ‘Structure of Feeling’ and the Problem of Democratic Values in Britain, 1938–1961,” *Modern Intellectual History* 17, no. 4 (2020): 1148. On the relationship between conventions and structures of feeling, and for a more granular account of the evolution of the latter concept, see David Simpson, “Raymond Williams: Feeling for Structures, Voicing History,” in *Cultural Materialism: On Raymond Williams*, ed. Christopher Prendergast (Minneapolis: University of Minnesota Press, 1995), 29–50.

15. Williams, *Marxism and Literature*, 130, 132.

16. Lawrence Grossberg, “Raymond Williams and the Absent Modernity,” in *About Raymond Williams*, ed. Monika Seidl, Roman Horak, and Lawrence Grossberg (New York: Routledge, 2010), 32.

17. See Clive Cazeaux, “Phenomenology and Radio Drama,” *British Journal of Aesthetics* 45, no. 2 (2005): 157–74.

18. Jason Loviglio, “The Traffic in Feelings: The Car-Radio Assemblage,” in *The Routledge Companion to Radio and Podcast Studies*, ed. Mia Lindgren and Jason Loviglio (New York: Routledge, 2022), 30.

19. Michel Foucault, *The Order of Things* (New York: Routledge, 2002); Robert Redfield, *Peasant Society and Culture: An Anthropological Approach to Civilization* (Chicago: University of Chicago Press, 1956).

20. Siobhán McHugh, "How Podcasting Is Changing the Audio Storytelling Genre," *Radio Journal: International Studies in Broadcast & Audio Media* 14, no. 1 (2016): 65–82. See also Benjamin Wallace-Wells, "The Strange Intimacy of *Serial*," *Vulture*, November 23, 2014, https://www.vulture.com/2014/11/strange-intimacy-of-serial.html; and Eugenia Williamson, "Oh, the Pathos!," *The Baffler* 20 (July 2012), https://thebaffler.com/salvos/oh-the-pathos.

21. John Biewen, "Introduction," in *Reality Radio: Telling True Stories in Sound*, 1st ed., ed. John Biewen and Alexa Dilworth (Chapel Hill: University of North Carolina Press, 2010), 5. The reader is advised to consult, as well, the 2017 second edition of this landmark series of essays by public radio producers, under the same title and publisher, which provides a very substantive update reflecting the rise of podcasting.

22. See Jason Loviglio, *Radio's Intimate Public: Network Broadcasting and Mass-Mediated Democracy* (Minneapolis: University of Minnesota Press, 2005); Bruce Lenthall, *Radio's America: The Great Depression and the Rise of Modern Mass Media Culture* (Chicago: University of Chicago Press, 2007); Neil Verma, "From Statement to Purpose: An Interview with Bill Siemering," *RadioDoc Review* 8, no. 1 (2022). See also the discussion of intimacy in chapter 1, this volume.

23. Alex Blumberg, "Power Your Podcast with Storytelling," CreativeLive, 2014, https://www.creativelive.com/class/power-your-podcast-storytelling-alex-blumberg.

24. Peabody Committee, "In the Dark" (APM Reports), Peabody Awards, 2017, http://www.peabodyawards.com/award-profile/in-the-dark.

25. Ellen McCracken, "The *Serial* Commodity: Rhetoric, Recombination, and Indeterminacy in the Digital Age," in *The Serial Podcast and Storytelling in the Digital Age*, ed. Ellen McCracken (New York: Routledge, 2017), 54–71.

26. Neil Verma, *Theater of the Mind: Imagination, Aesthetics and American Radio Drama* (Chicago: University of Chicago Press, 2012), 33–56.

27. See Scott Carrier, "That Jackie Kennedy Moment," in *Reality Radio: Telling True Stories in Sound*, 1st ed., ed. John Biewen and Alexa Dilworth (Chapel Hill: University of North Carolina Press, 2010), 35.

28. Lewis Raven Wallace, *The View from Somewhere: Undoing the Myth of Journalistic Objectivity* (Chicago: University of Chicago Press, 2019). See also Robert S. Boynton, "The View from Somewhere: A Review," *RadioDoc Review* 6, no. 1 (2020).

29. Ryan Engley, "The Impossible Ethics of *Serial*: Sarah Koenig, Foucault, Lacan," in *The "Serial" Podcast and Storytelling in the Digital Age*, ed. Ellen McCracken (New York: Routledge, 2017), 87–100.

30. Jason Loviglio, "*Criminal*: Journalistic Rigour, Gothic Tales and Philosophical Heft," *RadioDoc Review* 3, no. 1 (2017): 2.

31. Rebecca Ora, "Invisible Evidence: *Serial* and the New Unknowability of Documentary," in *Podcasting: New Aural Cultures and Digital Media*, ed. Dario Llinares, Neil Fox, and Richard Berry (New York: Palgrave, 2018), 107–22.

32. Michael Buozis, "Giving Voice to the Accused: *Serial* and the Critical Poten-

tial of True Crime," *Communication and Critical/Cultural Studies* 14, no. 3 (2017): 254–70.

33. Siobhán McHugh, *The Power of Podcasting: Telling Stories through Sound* (Sydney: University of New South Wales Press, 2022), 77.

34. John Biewen, "Little War on the Prairie: An Auto-Critique," *RadioDoc Review* 2, no. 1 (2015), https://ro.uow.edu.au/rdr/vol2/iss1/7/.

35. James T. Green, "Glass Walls," updated February 28, 2021, https://www.jamestgreen.com/thoughts/115. On how these critiques grew among emerging creators, see Jess Shane, "Toward a Third Podcasting: Activist Podcasting in an Age of Social Justice Capitalism," *RadioDoc Review* 8, no. 1 (2022).

36. For this episode I have relied on a translation on the *Radio Ambulante* site: "Translation—No Country for Young Men," *Radio Ambulante*, updated February 27, 2018, https://radioambulante.org/en/translation/translation-no-country-for-young-men.

37. Sarah Larson, "The Mystery of the Inconclusive Podcast," *New Yorker*, November 21, 2018, https://www.newyorker.com/culture/podcast-dept/the-mystery-of-the-inconclusive-podcast.

38. Rose Minutaglio, "April Balascio Suspected Her Father, Edward Wayne Edwards, Was a Serial Killer," *Elle*, July 18, 2019, https://www.elle.com/culture/movies-tv/a28396659/april-balascio-interview-the-clearing-podcast/.

39. Neroli Price, "Can True Crime Podcasts Make Structural Violence Audible?," in *The Routledge Companion to Radio and Podcast Studies*, ed. Mia Lindgren and Jason Loviglio (New York: Routledge, 2022), 358–67.

40. Stacey Copeland and Lauren Knight, "Indigenizing the National Broadcast Soundscape—CBC Podcast: *Missing and Murdered: Finding Cleo*," *Radio Journal* 19, no. 1 (2021): 101–16.

41. Jay Allison, "Introduction," in *Radio Diaries: DIY Handbook*, by Joe Richman and Jay Allison (Radio Diaries, 2016), viii.

42. Mia Lindgren, "Balancing Personal Trauma, Storytelling and Journalistic Ethics: A Critical Analysis of Kirsti Melville's *The Storm*," *RadioDoc Review* 2, no. 2 (2015): 1–15.

43. Seán Street, *The Poetry of Radio: The Colour of Sound* (New York: Routledge, 2013), 80.

44. Joe Richman and Jay Allison, *Radio Diaries: DIY Handbook* (Radio Diaries, 2016), 5–19; Joe Richman, "Diaries and Detritus: One Perfectionist's Search for Imperfection," in *Reality Radio: Telling True Stories in Sound*, 1st ed., ed. John Biewen and Alexa Dilworth (Chapel Hill: University of North Carolina Press, 2010), 128–29. See also Roland Barthes, *Camera Lucida: Reflections on Photography*, trans. Richard Howard (New York: Hill & Wang, 2010), 25–27.

45. Richman and Allison, *Radio Diaries*, 10.

46. Britta Jorgensen, "The Feelings Frontier: A Review of *No Feeling Is Final*," *RadioDoc Review* 5, no. 1 (2019), https://ro.uow.edu.au/rdr/vol5/iss1/4/.

47. Jennifer Lynn Stoever, *The Sonic Color Line: Race and the Cultural Politics of Listening* (New York: NYU Press, 2016).

48. Saidiya Hartman, "Venus in Two Acts," *Small Axe: A Caribbean Journal of Criticism* 12, no. 2 (2008): 1–14.

49. See Justin Eckstein, "Yellow Rain: Radiolab and the Acoustics of Strategic Maneuvering," *Journal of Argumentation in Context* 3, no. 1 (2014): 35–56; Paul Hillmer and Mary Ann Yang, "Commentary: Ignorance as Bias: *Radiolab*, Yellow Rain, and 'The Fact of the Matter,'" *Hmong Studies Journal* 18 (2017): 1–13.

50. James Phelan, "Narrative Progression," in *Narrative Dynamics: Essays on Time, Plot, Closure, and Frames*, ed. Brian Richardson (Columbus: The Ohio State University Press, 2002), 211, 214.

51. Sandra Kumamoto Stanley associates this aspect of true crime with postmodernism in an excellent early essay on *Serial*. Sandra Kumamoto Stanley, "'What We Know': Convicting Narratives in NPR's Serial," in *The "Serial" Podcast and Storytelling in the Digital Age*, ed. Ellen McCracken (New York: Routledge, 2017), 72–86.

52. Mark Seltzer, *True Crime: Observations on Violence and Modernity* (New York: Routledge, 2007), 15–16.

53. Kathleen Battles, *Calling All Cars: Radio Dragnets and the Technology of Policing* (Minneapolis: University of Minnesota Press, 2010).

54. Battles, *Calling All Cars*, 58.

55. Kathleen Battles and Amanda Keeler, "True Crime and Audio Media," in *The Routledge Companion to Radio and Podcast Studies*, ed. Mia Lindgren and Jason Loviglio (New York: Routledge, 2022), 191.

56. Claudia Calhoun, *Only the Names Have Been Changed: Dragnet, the Police Procedural and Postwar Culture* (Austin: University of Texas Press, 2022).

57. Caleb Kelly, *Cracked Media: The Sound of Malfunction* (Cambridge, MA: MIT Press, 2009).

58. Carlo Ginzburg, *Clues, Myths, and the Historical Method*, trans. John and Anne C. Tedeschi (Baltimore: Johns Hopkins University Press, 1992), 87–113.

59. See Rita Felski, "Suspicious Minds," *Poetics Today* 32, no. 2 (2011): 215–34. See also Alison Scott-Baumann, *Ricoeur and the Hermeneutics of Suspicion* (New York: Bloomsbury, 2009).

60. Jacob Smith, *Eco-Sonic Media* (Berkeley: University of California Press, 2015).

61. Erich Auerbach, *Mimesis: The Representation of Reality in Western Literature*, trans. Willard R. Trask, rev. ed., (Princeton: Princeton University Press, 2003), 6.

62. Auerbach, *Mimesis*, 23.

63. See, e.g., Susan J. Douglas, *Listening In: Radio and the American Imagination* (Minneapolis: University of Minnesota Press, 1999), 40–54; Douglas Kahn, *Earth Sound Earth Signal: Energies and Earth Magnitude in the Arts* (Berkeley: University of California Press, 2013), 25–82; Velimir Khlebnikov, "The Radio of the Future," in *Radiotext*, ed. Neil Strauss and David Mandl (New York: Autonomedia, 1993) (aka *Semiotext(e)* 16, vol. 6, no. 1 [1993]: 32–35); Sir Oliver Lodge, *Ether & Reality* (New York: George Doran,1925); Upton Sinclair, *Mental Radio* (Charlottesville, VA: Hampton Roads Books, 2001); Susan Merrill Squier, "Wireless Possibilities, Posthuman Possibilities: Brain Radio, Community Radio, Radio Lazarus," in *Communities of the Air: Radio Century, Radio Culture*, ed. Susan Merrill Squier (Durham: Duke University Press, 2003), 275–304; Gregory Whitehead, "Out of the Dark: Notes on the Nobodies of Radio Art," in *Wireless Imagination: Sound, Radio, and the*

Avant-Garde, ed. Douglas Kahn and Gregory Whitehead (Cambridge, MA: MIT Press, 1992), 253–63.

64. Jessica Abel, *Out on the Wire: The Storytelling Secrets of the New Masters of Radio* (New York: Broadway Books, 2015), 139–43.

65. This idea was spelled out by Hilmes in two articles released around the same time: Michele Hilmes, "The New Materiality of Radio: Sound on Screens," in *Radio's New Wave*, ed. Jason Loviglio and Michele Hilmes (New York: Routledge, 2013), 43–61; Michele Hilmes, "On a Screen Near You: The New Soundwork Industry," *Cinema Journal* 52, no. 3 (2013): 177–82. Subsequent authors whose approaches are rooted in this insight include Christopher Cwynar, "NPR Music: Remediation, Curation, and National Public Radio in the Digital Convergence Era," *Media, Culture & Society* 39, no. 5 (2017): 680–96; Amy Genders, "Radio as a Screen Medium in BBC Arts Broadcasting," *Journal of Radio & Audio Media* 25, no. 1 (2018): 142–55; and Andreas Lenander Aegidius, "How Radio Is Remediated in Streaming: The Case of Radio in Spotify," in *The Routledge Companion to Radio and Podcast Studies*, ed. Mia Lindgren and Jason Loviglio (New York: Routledge, 2022), 438–47.

66. Michael McDowell, "Pro Tools Proficiency May Be Keeping Us from Diversifying Audio," The Verge, January 22, 2021, https://www.theverge.com/2021/1/22/22242606/pro-tools-proficiency-podcasting-diversity-gatekeeping. See also Mike D'Errico, *Push: Software Design and the Cultural Politics of Music Production* (New York: Oxford, 2022), 51–84.

67. Katherine Quanz, "Pro Tools, Playback, and the Value of Postproduction Sound Labor in Canada," *Velvet Light Trap* 76 (2015): 37–48.

68. Jeremy Lakoff, "Maestro, If You Please: The Radio Producer as Musician," in *Radio Art and Music: Culture, Aesthetics, Politics*, ed. Jarmila Mildorf and Pim Verhulst (New York: Lexington Books, 2020), 37–39.

69. "Using Reaper for Radio and Podcasting," YouTube, 3:48:40, March 5, 2021, https://www.youtube.com/watch?v=xbyPl6YjiVI.

70. Jonathan Sterne, *The Audible Past: Cultural Origins of Sound Reproduction* (Durham: Duke University Press, 2003), 14–15.

71. See, e.g., Michel Foucault, *Power/Knowledge: Selected Interviews and Other Writings: 1972–1977*, ed. Colin Gordon, trans. Colin Gordon, Leo Marshall, John Mepham, and Kate Soper (New York: Vintage Books, 1980), 119.

Chapter 3

1. Maya Dukmasova, "Radio Gaga," *Chicago Reader* 47, no. 11 (December 14, 2017): 14–22.

2. "About," The Sarahs, accessed January 22, 2023, http://thesarahawards.com/about.

3. See, e.g., Nicholas Quah, "11 Great Podcast Recommendations for Fall 2017," *Vulture*, September 28, 2017, https://www.vulture.com/2017/09/11-great-podcast-recommendations-fall-2017.html; Michelle Harven, "Fiction Podcasts Are on the Rise," *Washington Post*, May 23, 2019, https://www.washingtonpost.com/entertainment/books/fiction-podcasts-are-on-the-rise-and-better-than-ever-the

se-are-the-ones-to-listen-to/2019/05/22/7afcc2ae-7714–11e9-b7ae-390de42596 61_story.html.

4. Kevin L. Jones, "EARFUL: Talking with Jonathan Mitchell of 'The Truth,'" KQED, September 1, 2015, https://www.kqed.org/arts/10928068/earful-talking-with-jonathan-mitchell-of-the-truth.

5. Quoted in the forum section for a series hosted by radio artist Joan Schuman's audio website, "Radio's Art," Earlid, last modified September 5, 2017, https://www.earlid.org/posts/radios-art/.

6. Sarah Larson, "At Third Coast, Holding on to Public-Radio Ideals as Podcasting Booms," *New Yorker*, November 18, 2017.

7. See, e.g., Jack Smart, "Radio Drama: The New Frontier," Backstage, last updated August 5, 2019, https://www.backstage.com/magazine/article/radio-drama-new-frontier-9058/.

8. See Hugh Chignell, *British Radio Drama, 1945–63* (London: Bloomsbury, 2019); Mark E. Cory, "Soundplay: The Polyphonous Tradition of German Radio Art," in *Wireless Imagination: Sound, Radio and the Avant-Garde*, ed. Douglas Kahn and Gregory Whitehead (Cambridge, MA: MIT Press, 1992), 332–71; Liz Gunner, *Radio Soundings: South Africa and the Back Modern* (London: Cambridge University Press, 2019); Bruce Lenthall, *Radio's America: The Great Depression and the Rise of Modern Mass Culture* (Chicago: University of Chicago Press, 2007); Tom McEnaney, *Acoustic Properties: Radio, Narrative, and the New Neighborhood of the Americas* (Evanston, IL: Northwestern University Press, 2017), 140–66; Neil Verma, *Theater of the Mind: Imagination, Aesthetics and American Radio Drama* (Chicago: University of Chicago Press, 2012); and Sonja D. Williams, *Word Warrior: Richard Durham, Radio, and Freedom* (Urbana: University of Illinois Press, 2015). For students new to the historical study of radio drama, see also Lars Bernaerts and Jarmila Mildorf, eds., *Audionarratology: Lessons from Radio Drama* (Columbus: The Ohio State University Press, 2021); Debra Rae Cohen, Michael Coyle, and Jane Lewty, eds., *Broadcasting Modernism* (Gainesville: University Press of Florida, 2009); Tim Crook, *Radio Drama: Theory and Practice* (New York: Routledge, 1999); Daniel Gilfillan, *Pieces of Sound: German Experimental Radio* (Minneapolis: University of Minnesota Press, 2009); Elissa Guralnick, *Sight Unseen: Beckett, Pinter, Stoppard, and Other Contemporary Dramatists on Radio* (Athens: Ohio University Press, 1995); Richard J. Hand, *Terror on the Air! Horror Radio in America, 1931–1952* (Jefferson, NC: Mcfarland, 2006); Michele Hilmes and Jason Loviglio, eds., *Radio Reader: Essays in the Cultural History of Radio* (New York: Routledge, 2002); Leslie McMurtry, *Revolution in the Echo Chamber: Audio Drama's Past, Present and Future* (Chicago: Intellect, 2021); Jeff Porter, *Lost Sound: The Forgotten Art of Radio Storytelling* (Chapel Hill: University of North Carolina Press, 2016); Elena Razlogova, *The Listener's Voice: Early Radio and the American Public* (Philadelphia: University of Pennsylvania Press, 2011); Alexander Russo, *Points on the Dial: Golden Age Radio beyond the Networks* (Durham: Duke University Press, 2010); Shawn VanCour, *Making Radio: Early Radio Production and the Rise of Modern Sound Culture* (New York: Oxford University Press, 2018).

9. David Bordwell, *Reinventing Hollywood: How 1940s Filmmakers Changed Movie Storytelling* (Chicago: University of Chicago Press, 2017), 6–7.

10. Andrew J. Bottomley, "Podcasting, *Welcome to Night Vale*, and the Revival of Radio Drama," *Journal of Radio & Audio Media* 22, no. 2 (2015): 179–89.

11. Jacques Derrida, *Archive Fever: A Freudian Impression*, trans. Eric Prenowitz (Chicago: University of Chicago Press, 1996).

12. Michele Hilmes, "The Lost Critical History of Radio," *Australian Journalism Review* 36, no. 2 (2014): 11–22.

13. Thomas Elsaesser, "Freud as Media Theorist: Mystic Writing-Pads and the Matter of Memory," *Screen* 50, no. 1 (2009): 107; Gilles Deleuze, *Cinema 2: The Time Image*, trans. Hugh Tomlinson and Robert Galeta (Minneapolis: University of Minnesota Press, 1989), 56.

14. The relevant passages are two connected texts: Sigmund Freud, *Beyond the Pleasure Principle*, trans. James Strachey (New York: W. W. Norton, 1961) 31–39; and Sigmund Freud, "A Note Upon the 'Mystic Writing-Pad' (1925)," in *The Standard Edition of the Complete Psychological Works of Sigmund Freud*, vol. 19, ed. and trans. James Strachey (London: Hogarth Press, 1961).

15. Andreas Huyssen, *Present Pasts: Urban Palimpsests and the Politics of Memory* (Stanford, CA: Stanford University Press, 2003).

16. See Sallie Baxendale, "Memories Aren't Made of This: Amnesia at the Movies," *British Medical Journal* 329, no. 7480 (2004): 1480–83.

17. Crook, *Radio Drama*, 21.

18. Paul Raeburn, "Just Like the Old Days, Except for the .com," *New York Times*, October 25, 1998, http://www.nytimes.com/1998/10/25/arts/television-radio-just-like-the-old-days-except-for-the-com.html.

19. Wil Williams, "I Don't Care What We Call Non-Non-Fiction Podcasts," Wil Williams Writes for Themself, June 12, 2019, https://wilwilliams.reviews/2019/06/12/i-dont-care-what-we-call-non-non-fiction-podcasts/.

20. D. N. Rodowick, *What Philosophy Wants from Images* (Chicago: University of Chicago Press, 2017), 4.

21. This scenario was oddly prefigured by Alan Weiss. See Alan S. Weiss, *Breathless: Sound Recording, Disembodiment, and the Transformation of Lyrical Nostalgia* (Middletown, CT: Wesleyan University Press, 2002), xi.

22. Martin Spinelli and Lance Dann, *Podcasting: The Audio Media Revolution* (New York: Bloomsbury, 2019), 124.

23. Ella Watts, "Queer Networks versus Global Corporations: The Battle for the Soul of Audio Fiction," in *The Bloomsbury Handbook of Radio*, ed. Kathryn McDonald and Hugh Chignell (London: Bloomsbury, 2023), 438–39.

24. McMurtry, *Revolution in the Echo Chamber*, 36–37.

25. Kc Wayland, *Bombs Always Beep: Creating Modern Audio Theater* (Orange, CA: Wayland Productions, 2018).

26. Wil Williams and Tal Minear, "The Editors' Response to Rusty Quill's Statement," Medium, January 26, 2023, https://talminear.medium.com/the-editors-response-to-rusty-quill-s-statement-57c9485f9ac2. Minear currently distributes a newsletter called *Podplane* centered on trans and nonbinary podcasts.

27. Alexandra Alter, "An Art Form Rises: Audio without the Book," *New York Times*, November 30, 2014, https://www.nytimes.com/2014/12/01/business/media/new-art-form-rises-audio-without-the-book-.html; Caitlin Flynn, "X-Philes, Get

Excited about This New Audiobook," Refinery29, April 11, 2017, https://www.refi nery29.com/en-gb/2017/04/149452/xfiles-audiobook-gillian-anderson.

28. For a fascinating and illuminating account of this aesthetic, see Farokh Soltani, *Radio/Body: Phenomenology and Dramaturgies of Radio* (Manchester, UK: Manchester University Press, 2020), 191–210.

29. Peter Marks, "Now Hear This! Playwrights Are Making Theater for the Ears. Yes, That's a Thing," *Washington Post*, November 26, 2019, https://www.washi ngtonpost.com/entertainment/theater_dance/now-hear-this-playwrights-are-mak ing-theater-for-the-ears-yes-thats-a-thing/2019/11/26/76bfb9a2–0ed7–11ea-b0fc -62cc38411ebb_story.html.

30. McMurtry, *Revolution in the Echo Chamber*, 16.

31. Verma, *Theater of the Mind*, 4–5.

32. For typical examples in the United States, see Erik Barnouw, *Handbook of Radio Writing* (New York: Little, Brown, 1947), 27–80; and A. H. Lass, Earle McGill, and Donald Axelrod, *Plays from Radio* (New York: Houghton Mifflin, 1948), 4–5. For the classic example from the United Kingdom, see Donald McWhinnie, *The Art of Radio* (London: Faber & Faber, 1959), 21–102. For a much broader ranging and deeper understanding of how dramatists thought about their "ingredients" over time, the essential text is Tim Crook, "The Audio Dramatist's Critical Vocabulary in Great Britain," in *Audionarratology: Lessons from Radio Drama*, ed. Lars Bernaerts and Jarmila Mildorf (Columbus: The Ohio State University Press, 2021), 19–50. Another key book that takes on the same questions from a theoretical point of view is Andrew Crisell, *Understanding Radio* (New York: Routledge, 1996), 42–66.

33. The theory of recording vocalized poetry and literature is its own field that grapples with some of these issues, only from the other direction. On this, see, e.g., Jason Camlot, *Phono Poetics: The Making of Early Literary Recordings* (Stanford, CA: Stanford University Press, 2019); Charles Bernstein, ed., *Close Listening: Poetry and the Performed Word* (New York: Oxford University Press, 1998); Marit J. MacArthur, "Monotony, the Churches of Poetry Reading, and Sound Studies," *PMLA* 131, no. 1 (2016): 38–63; and Matthew Rubery, ed., *Audiobooks, Literature, and Sound Studies* (New York: Routledge, 2011).

34. Milton Allen Kaplan, *Radio and Poetry* (New York: Columbia University Press, 1949), 110.

35. Sarah Montague, "Towards a Poetics of Audio: The Importance of Criticism," The Sarahs, April 3, 2017, http://thesarahawards.com/article/2017/4/3/towa rds-a-poetics-of-audio-the-importance-of-criticism.

36. Robert J. Landry, "The Improbability of Radio Criticism," *Hollywood Quarterly* 2, no. 1 (1946): 66–70.

37. Remy Brunel, "Radio Magic Expected to Emerge from Columbia Workshop Experimentation," *Washington Post*, September 6, 1936.

38. Earle McGill, *Radio Directing* (New York: McGraw-Hill, 1940).

39. Amanda Hess, "The Story So Far: Fiction Podcasts Take Their Next Steps," *New York Times*, November 11, 2016, https://www.nytimes.com/2016/11/13/arts/fi ction-podcasts-homecoming.html; Joanna Robinson, "Oscar Isaac and Catherine Keener Present Your New Podcast Obsession," *Vanity Fair*, November 16, 2016, http://www.vanityfair.com/hollywood/2016/11/homecoming-podcast-oscar-isaac -catherine-keener-david-schwimmer-gimlet.

40. Eleanor Patterson, "Reconfiguring Radio Drama after Television: The Historical Significance of *Theater 5*, *Earplay* and *CBS Radio Mystery Theater* as Post-Network Radio Drama," *Historical Journal of Film, Radio and Television* 36, no. 4 (2016): 649–67.

41. Robinson, "Oscar Isaac and Catherine Keener."

42. Archibald MacLeish, *The American Story: Ten Broadcasts* (New York: Duell, Sloan and Pearce, 1944), xi.

43. Quoted in John Drakakis, ed., *British Radio Drama* (Cambridge: Cambridge University Press, 1981), 11.

44. Yuri Rasovsky, *The Well-Tempered Radio Dramatist* (Hamsted, NY: National Audio Theater Festivals, 2006), 11.

45. See the recent special section on radio drama from *RadioDoc Review*, featuring the work of Emily Lane, Hugh Chignell, and Leslie McMurtry: Leslie McMurtry, "Special Section: Radio Drama Takeover," *RadioDoc Review* 7, no. 1 (2021), https://ro.uow.edu.au/rdr/vol7/iss1/4/.

46. See James Naremore, *More Than Night: Film Noir in Its Contexts* (Berkeley: University of California Press, 1998).

47. See Michael C. Keith and Mary Ann Watson, eds., *Norman Corwin's One World Flight: The Lost Journal of Radio's Greatest Writer* (New York: Continuum, 2009); Jacob Smith, "Norman Corwin's Radio Realism," in *Anatomy of Sound: Norman Corwin and Media Authorship*, ed. Jacob Smith and Neil Verma (Oakland: University of California Press, 2016), 101–26; Jacob Smith, "Travels with Jack: ZBS's Post-Network Radio Adventure," *Resonance* 1, no. 1 (2020): 94–115. For even earlier uses of wire recording in narrative radio, see Ian Whittington, *Writing the Radio War: Literature, Politics and the BBC, 1939–1945* (Edinburgh: Edinburgh University Press, 2018), 117–52.

48. Verma, *Theater of the Mind.*

49. I credit this idea to Farokh Soltani's response to an earlier published version of this chapter. Soltani suggests that one place to look for a future sound of audio drama might be to dramaturgical approaches that did not survive the experimental era. See Farokh Soltani, "Inner Ears and Distant Worlds: Podcast Dramaturgy and the *Theatre of the Mind*," in *Podcasting: New Aural Cultures and Digital Media*, ed. Dario Llinares, Neil Fox, and Richard Berry (New York: Palgrave McMillan, 2018), 189–208.

50. Donald McWhinnie, "Introduction," in *Six Plays for Radio*, by Giles Cooper (London: BBC, 1996), 10.

51. Richard J. Hand and Mary Traynor, *The Radio Drama Handbook* (New York: Continuum, 2011), 56.

52. Earle McGill, *Radio Directing* (New York: McGraw-Hill, 1940), 12.

53. See also Andrew Leland, "The Radio Auteur: Joe Frank, Ira Glass, and Narrative Radio," *New York Review of Books*, September 29, 2018, https://www.nybooks.com/online/2018/09/29/the-radio-auteur-joe-frank-ira-glass-and-the-legacy-of-narrative-radio/.

54. Here I am influenced by Derrida's argument about how the mnemic and the anamnestic—modes of memory present in the self and the body—are always already defined by hypomnesis, the knowledge we encode, externalize, and stow

away in places of consignation. See Derrida, *Archive Fever*, 11–20; Bernard Stiegler, *What Makes Life Worth Living: On Pharmacology*, trans. Daniel Ross (Malden, MA: Polity Press, 2013).

55. David Hendy, *Life on Air: A History of Radio Four* (New York: Oxford University Press, 2007), 190–94.

56. Erik Barnouw, *The Radio Writing Handbook* (Boston: Little, Brown, 1947), 23.

57. For more on this, see Clive Cazeaux, "Phenomenology and Radio Drama," *British Journal of Aesthetics* 45, no. 2 (2005): 157–74.

58. Mary Louise Hill, "Developing *A Blind Understanding*: A Feminist Revision of Radio Semiotics," in *Experimental Sound and Radio*, ed. Allen S. Weiss (Cambridge, MA: MIT Press, 2001), 107–15.

59. Elena Fernández Collins, "A Collection of Beloved Tapes: *The Bright Sessions* Retrospective," Bello Collective, July 13, 2018, https://bellocollective.com/a-collection-of-beloved-tapes-the-bright-sessions-retrospective-a412719358c9; Emma Dibdin, "6 Podcasts for the Drama Lover," *New York Times*, August 8, 2019, https://www.nytimes.com/2019/08/15/arts/podcast-dramas.html.

60. Rudolf Arnheim, *Radio*, trans. Margaret Hedwig and Herbert Read (London: Faber and Faber, 1971), 125.

61. Marie-Laure Ryan, "Stacks, Frames and Boundaries, or Narrative as Computer Language," in *Narrative Dynamics: Essays on Time, Plot, Closure, and Frames*, ed. Brian Richardson (Columbus: The Ohio State University Press, 2002), 366–86. I am grateful to have been reminded of this approach by reading the recent work of Caroline Kita. See Caroline Kita, "Simultaneity and the Soundscapes of Audio Fiction," in *Audionarratology: Lessons from Radio Drama*, ed. Lars Bernaerts and Jarmila Mildorf (Columbus: The Ohio State University Press, 2021) 105–6.

62. Verma, *Theater of the Mind*, 91–133.

63. Lauren Berlant, *Cruel Optimism* (Durham: Duke University Press, 2012), 24–25.

64. Jack Halberstam, *In a Queer Time and Place: Transgender Bodies, Subcultural Lives* (New York: NYU Press, 2005). This title may also be found under the byline Judith Halberstam.

65. Mieke Bal, *Narratology: Introduction to the Theory of Narrative*, 2nd ed. (Toronto: University of Toronto Press, 1997), 90.

66. Roland Barthes, "The Grain of the Voice," in *Image-Music-Text*, trans. Stephen Heath (New York: Noonday Press, 1977), 179–89.

67. Gregory Whitehead, "Radio Play Is No Place," in *Experimental Sound & Radio*, ed. Allen S. Weiss (Cambridge, MA: MIT Press, 2001), 89–94.

68. Sara Ahmed, *Queer Phenomenology: Orientations, Objects, Others* (Durham: Duke University Press, 2006). 27.

69. Ahmed, *Queer Phenomenology*, 54.

70. Svetlana Boym, *The Future of Nostalgia* (New York: Basic Books, 2001), 8.

71. Elena Fernández Collins, "This Is Not a Profile about Kaitlin Prest," Bello Collective, November 1, 2019, https://bellocollective.com/this-is-not-a-profile-about-kaitlin-prest-57437ff1836e.

72. Michael Paulson, "Radio Drama for a Podcast Age: How Amazon's Audible

Moved into Theater," *New York Times*, March 9, 2022, https://www.nytimes.com/2022/03/09/theater/amazon-audible-theater.html; Christina Goldbaum, "Miss the N.Y.C. Subway? These Radio Plays Bring It Back to Life," *New York Times*, December 24, 2020, https://www.nytimes.com/2020/12/24/theater/mta-plays-subway-rattlestick.html; Arifa Akbar, "The Lockdown Boom in Audio Plays," *The Guardian*, February 2, 2021, https://www.theguardian.com/stage/2021/feb/02/the-one-form-of-drama-the-pandemic-cant-touch-the-rise-of-audio-plays; Jeff Lunden, "Theaters Return to an Old Art Form—the Radio Drama—with a Twist," NPR, June 9, 2020, https://www.npr.org/2020/06/09/865061847/theaters-return-to-an-old-art-form-the-radio-drama-with-a-twist. For transparency, I would like to point out that I was one of the quoted sources in the Paulson article.

73. Rashika Rao, "The Left-Right Game Proves Not Everything Should Be a Podcast," Medium, October 6, 2020, https://bellocollective.com/qcode-medias-sci-fi-fiction-the-left-right-game-proves-not-everything-should-be-a-podcast-f3a8e5903c34.

74. On the web series example, from a generation earlier, see Aymar Jean Christian, *Open TV: Innovation beyond Hollywood and the Rise of Web Television* (New York: NYU Press, 2018).

75. Kate Arthur, "How Podcast Leader QCode Uses Star Power," *Variety*, October 22, 2021, https://variety.com/2021/biz/news/qcode-podcast-explosion-1235093655/.

76. Reggie Ugwu, "'Podcast Movies'? Feature-Length Fiction Stretches the Medium," *New York Times*, December 24, 2021, https://www.nytimes.com/2021/12/24/arts/podcast-movies-fiction.html.

77. Joshua Dudley, "What Do We Call Feature-Length Podcasts?," *Forbes*, December 31, 2021, https://www.forbes.com/sites/joshuadudley/2021/12/31/what-do-we-call-feature-length-fiction-podcasts/?sh=353d15012af1.

Coda

1. The broadcast is available online: DinahBird and Jean-Philippe Renoult, "The Gold Line," YouTube, 19:02, May 15, 2020, https://www.youtube.com/watch?v=AyhxbZ77jPI.

2. The Radius archive is available at theradius.us.

3. Gregory Whitehead, "Out of the Dark: Notes on the Nobodies of Radio Art," in *Wireless Imagination, Sound Radio, and the Avant Garde*, ed. Doublas Kahn and Gregory Whitehead (Cambridge, MA: MIT Press, 1992), 253.

4. See, e.g., Tetsuo Kogawa, "Free Radio in Japan: The Mini FM Boom," in Neil Strauss, ed., "Radiotext(e)," *Semiotext(e)* 16, vol. 6, no. 1 (1993): 85–90. For more on the roots of contemporary radio art, see, e.g., Knut Aufermann, Helen Hahmann, Sara Washington, and Ralf Wendt, *Radio Revolten: 30 Days of Radio Art* (Leipzig: Spector Books, 2019); Colin Black, "Contextualising Australian Radio Art," *Radio Journal* 18, no. 2 (2017): 271–89; Daniel Gilfillan, *Pieces of Sound: German Experimental Radio* (Minneapolis: University of Minnesota Press, 2009); Magz Hall, "The New Wave: On Radio Arts in the UK," *Sounding Out!*, December 11, 2014, https://soundstudiesblog.com/2014/12/11/the-new-wave-on-radio-arts-in

-the-uk/; Galen Joseph-Hunter, Penny Duff, and Maria Papadomanolaki, eds., *Transmission Arts: Artists and Airwaves* (New York: PAJ Publications, 2011); Douglas Kahn, *Earth Sound Earth Signal: Energies and Earth Magnitude in the Arts* (Berkeley: University of California Press, 2013); Douglas Kahn and Gregory Whitehead, eds., *Wireless Imagination, Sound Radio, and the Avant Garde* (Cambridge, MA: MIT Press, 1992); Ann Thurmann-Jajes and Regine Beyer, eds., *Listen Up! Radio Art in the USA* (New York: Columbia University Press, forthcoming); and Allen S. Weiss, *Experimental Sound and Radio* (Cambridge, MA: MIT Press, 2001).

5. Anna Friz, "Someplaces: Radio Art, Transmission Ecology and Chicago's 'Radius,'" *Sounding Out*, November 6, 2014, https://soundstudiesblog.com/2014/11/06/someplaces-radio-art-transmission-ecology-and-chicagos-radius/. The concept of transmission ecology has been central to Friz's thought and practice for some time. See also, e.g., Anna Friz, "Transmission Art in the Present Tense," *PAJ: A Journal of Performance and Art*, 31, no. 3 (2009): 46–49; Anna Friz, "The Radio of the Future Redux: Rethinking Transmission through Experiments in Radio Art" (PhD diss., York University, 2011); and Gascia Ouzounian, "Contemporary Radio Art and Spatial Politics: The Critical Radio Utopias of Anna Friz," *Radio Journal* 5, no. 2–3 (2008): 129–42.

6. John Lewis, "Yvette Janine Jackson: Freedom Review—Vivid Voyage through Hate," *The Guardian*, January 15, 2021, https://www.theguardian.com/music/2021/jan/15/yvette-janine-jackson-freedom-review.

7. There were, of course, exceptions in organizations that made heroic attempts to bring the worlds of radio art and podcasting together in this period. One from the podcasting side was the podcast *Constellations*, curated by Michelle Macklem and Jess Shane (https://constellationssounds.org/). One from the radio art side was Joan Schuman's Earlid project, which curated web-based radio art that was also broadcast on Wave Farm (https://www.earlid.org/).

8. The Wave Farm Broadcast Radio Art Archive is available at https://wavefarm.org/radio/archive/artists.

Index

Page numbers in *italics* refer to figures.